The Plumber's Troubleshooting Guide

D0776114

The Plumber's Troubleshooting Guide

R. Dodge Woodson

McGraw-Hill, Inc.

New York   San Francisco   Washington, D.C.   Auckland   Bogotá
Caracas   Lisbon   London   Madrid   Mexico City   Milan
Montreal   New Delhi   San Juan   Singapore
Sydney   Tokyo   Toronto

Library of Congress Cataloging-in-Publication Data

Woodson, R. Dodge (Roger Dodge)
    The plumber's troubleshooting guide / R. Dodge Woodson.
        p.   cm.
    Includes index.
    ISBN 0-07-071777-X
    1. Plumbing—Repairing.   I. Title.
    TH6681.W66   1994
    696'.1—dc20   94-4065
                                                          CIP

1 2 3 4 5 6 7 8 9 0   DOC/DOC   9 0 9 8 7 6 5 4

ISBN 0-07-071777-X

The sponsoring editor for this book was Larry Hager, the editing supervisor was Kimberly A. Goff, and the production supervisor was Donald F. Schmidt. This book was set in Century Schoolbook. It was composed by McGraw-Hill's Professional Book Group composition unit.

Printed and bound by R. R. Donnelley & Sons Company.

This book is dedicated to my daughter, Afton Amber Woodson, and my wife, Kimberley. Kimberley has been my best friend for twelve years, and Afton is my inspiration. These ladies of my life are my life.

Contents

Preface

Troubleshooting is one area of plumbing where you can never be too experienced. Regardless of how much you know and how many years you've been in the trade, there will always be some little problem that causes you grief.

I've been in the plumbing trade for about 20 years, and there are still times when I run across situations unlike any others I've ever seen. As I've supervised plumbers over the years, I've noticed that many of them lack specialized troubleshooting skills; that is what prompted me to write this book.

Most plumbers can eventually solve their problems in the field, but a lot of them waste a tremendous amount of time, energy, and money in doing so. With the help of this book, you can minimize your lost time and maximize your profit potential by getting to the root cause of problems faster.

What is this book going to do for you? It is going to show you a system of troubleshooting procedures that have been refined and perfected to a point where they are extremely effective and efficient. You will learn how to change the way you approach problems in the field, and you will be given step-by-step instructions for becoming a proficient troubleshooter.

Take a few moments to glance over the table of contents. You will see that this is a comprehensive field guide that can help you solve any type of plumbing problem you might encounter. Whether you are troubleshooting a bidet, a bedpan washer, dental equipment, or a common kitchen sink, this book gives you the steps needed to find fast, dependable solutions to problems.

What makes this book so good? There are many good attributes surrounding these pages. The most important is the value of the information contained between the covers of the book. It is information gathered over a long, successful career as a master plumber and plumbing contractor. Having the benefit of my 20 years of experience is enough to make this book special, but there's more. Unlike so many books on technical subjects, this book is easy to read and understand. There are hundreds of illustrations that show you detailed information on everything from faucet construction to friction loss. The chapters are laid out to make it easy for you to use only the portion of the book you need, when you need it. Go ahead, take a minute to thumb through the pages. I'm sure you'll see immediately that this is not your average off-the-shelf plumbing book.

Whether you are a seasoned master plumber, a journeyman, or an apprentice, you are sure to discover aspects of plumbing of which you were unaware.

Not only will you learn solid troubleshooting skills by reading this book, you will also be entertained with real-life case histories, such as the time I found a beer can in a toilet trap. There are plenty of true stories mixed into the text that teach as they entertain.

Buy this book, read it, and put it in your truck. You will be surprised at how often you will refer to it as a reference guide. No working plumber or plumbing contractor should be without the information contained herein.

R. Dodge Woodson

Acknowledgments

I would first like to acknowledge and thank my parents, Maralou and Woody, for always being there when they're needed.

The following companies are thanked for their art contributions to this book:

Amtrol, Inc.

A. O. Smith Water Products Co.

A. Y. McDonald Manufacturing Company

Crane Plumbing Fixtures

Delta Faucet Company

EBCO Manufacturing Company

Goulds Pumps, Inc.

Hellenbrand Water Conditioners, Inc.

Moen, Inc.

UNR Home Products

Symmons Industries, Inc.

Woodford Manufacturing Company

Introduction

Chapter 1: Understanding the Art of Troubleshooting. It takes time and experience to learn the art of troubleshooting, but once you do, you will work faster and more effectively. This chapter details all aspects of the principles involved with professional troubleshooting techniques.

Chapter 2: How Effective Troubleshooting Saves Time and Money. This chapter shows you why it is worthwhile to learn proper troubleshooting methods. You will see how good troubleshooting skills can save you both time and money, not to mention frustration.

Chapter 3: Troubleshooting Toilets. This chapter addresses all issues of troubleshooting toilets. Whether you are working with wall-hung toilets, floor-mounted toilets, toilets with flush tanks, or toilets with flush valves, this chapter shows you what to look for and in what order to eliminate possibilities.

Chapter 4: Troubleshooting Sinks and Lavatories. This chapter deals with all types of sinks—service sinks, kitchen sinks, laundry sinks, and others—and lavatories—wall-hung, pedestal, drop-in, rimmed, and handicap lavatories. You will learn what to look for and how to look for it when troubleshooting sinks and lavatories.

Chapter 5: Troubleshooting Bathtubs. This chapter concentrates on the problems associated with all types of bathtubs. You will learn how to check out tub wastes and overflows and how to find mystery leaks.

Chapter 6: Troubleshooting Showers. Whether you are dealing with a built-up shower, one-piece shower, sectional shower, or emergency shower, this chapter will answer your questions.

Chapter 7: Troubleshooting Spas and Whirlpools. Working with spas and whirlpools can get very complicated, but this chapter removes the confusion with which you may be faced. You will learn more about spas and whirlpools than you thought there was to know.

Chapter 8: Troubleshooting Bidets.   Most plumbers don't work with bidets on a daily basis, and this leads to a lack of experience in many plumbing professionals. Never fear, this chapter is going to clear up your questions on troubleshooting bidets.

Chapter 9: Troubleshooting Urinals.   When residential plumbers want to branch out into commercial work, they must acquaint themselves with new types of fixtures, one of which is an urinal. If you are not familiar with the workings of urinals, this chapter will be indispensable.

Chapter 10: Troubleshooting Faucets and Valves.   With all the faucets and valves available to today's plumbers, it can be extremely difficult to keep up with them all. This chapter gives you an in-depth rundown on all the most common types of faucets and valves. It will take a while to read it, but when you are finished, your knowledge will be greatly enhanced.

Chapter 11: Troubleshooting Plumbing Appliances.   Why isn't that ice-maker working? What's wrong with that garbage disposer? Where is that dishwasher leak coming from? Why isn't that water heater acting properly? This chapter will answer all these questions and many more on plumbing appliances.

Chapter 12: Troubleshooting Waste-Water Pumps.   It is not every day that plumbers are asked to troubleshoot waste-water pumps, but the calls do come in with some regularity. Do you know how to evaluate problems with waste-water pumps? If you don't, you will after reading this chapter.

Chapter 13: Troubleshooting Specialty Fixtures.   Some plumbers want to get into the lucrative field of hospital plumbing, and these plumbers need special skills. Learning to troubleshoot specialty fixtures can be a time-consuming process, but this chapter shortens the learning curve.

Chapter 14: Troubleshooting Well Systems.   Well systems and water pumps are like a foreign language to some plumbers. Plumbers moving from cities to rural locations have no knowledge of water pumps, but they are required to work with them. How can you gain a fast education in water pumps? You can read this chapter.

Chapter 15: Troubleshooting Septic Systems.   This chapter does for septic systems what Chap. 14 does for wells systems.

Chapter 16: Troubleshooting Water Treatment Systems.   Water treatment is becoming a big business. More and more people are having equipment installed to treat their water. Somebody has to service and repair this equipment. If you have the skills needed to troubleshoot water treatment systems, you have the power to make more money. This chapter shows you what is needed to work with most types of water treatment systems.

Chapter 17: Troubleshooting Water Distribution Systems. Many plumbers never consider that they may be asked to solve problems associated with water distribution systems. After all, it's just pipe. How hard could it be to troubleshoot? Do you know what would cause pinholes in copper tubing? Do you know why water pipes might bang when the water is used? This chapter is going to show you a few things about water pipes that you may not know.

Chapter 18: Troubleshooting Drainage and Vent Systems. What would you do if a customer asked you to explain where a terrible smell was coming from? Would you look for a blocked vent or a trap that had dried up from lack of use? If a drainage system is having recurring problems with stoppages, what would you look for? This chapter is going to show you what to expect from the drainage and vent systems in buildings.

1

Understanding the Art
of Troubleshooting

For plumbers, understanding the art of troubleshooting problems effectively is a blessing. It saves time, money, and frustration for both the plumbers and their customers. Good troubleshooting is much more than hunting randomly for the cause of a problem. Proficient troubleshooters don't waste time chasing after solutions that are not needed. They create a profile of the problem, assess the data available to them, and move on, in a concise process of elimination, to find the root cause of a problem.

This book is dedicated to teaching you how to develop keen troubleshooting skills. Most of the book deals with problems specific to plumbing, but this chapter can apply to any type of problem. Here, you will learn the foundation for building strong troubleshooting abilities.

I have used the techniques you are about to be given for many years, and I can tell you, they work. If you absorb the basics from this chapter, you can apply the principles to any plumbing problem you're faced with. In fact, you can use the same procedures to solve business problems, personal problems, or any other type of problem you may run into. Now, if you're ready, let's see what steps professional troubleshooters take to solve their problems effectively.

Effective Problem Solving

What is effective problem solving? It is the act of transforming problems into goals and meeting those goals. Effective problem solving provides sensible solutions to all types of problems, both personal and professional.

Benefits of Effective Problem Solving

Everyone can benefit from effective problem solving. The troubleshooting system you are about to learn can be applied to any type of problem, and the principles are sound for all aspects of problem solving and dedicated decision making.

Eight-Stage Problem-Solving System

My personal problem-solving system involves eight stages. If any of the stages are not used properly, the system will not work to its full potential. The eight stages are listed below:

- Realizing you have a problem
- Evaluating the effects of the problem
- Identifying the problem
- Developing potential solutions to the problem
- Assessing sensible solutions
- Picking a plan of attack and implementing it
- Tracking your results
- Making progressional adjustments

Each of these stages are critical in solving problems quickly, effectively, and positively. The eight stages, however, are only the skeleton of the system. For a problem-solving plan to work, it must be specific, explicit, efficient, and cost effective. My system gives you the power to create a winning plan for problem solving and dedicated decision making in all your plumbing work. Let's look at each of the eight stages a little more closely.

Realizing you have a problem

Realizing you have a problem to solve is the first step in implementing effective troubleshooting techniques. This step is not usually much of a consideration for service plumbers. When they get a call from a customer, they know there is some type of problem to be dealt with.

As a plumbing contractor, however, problems in your business may not be so apparent. It may seem that you would know automatically that you have a problem, but many people and businesses go from for months without realizing a problem exists. Problems that have existed for long periods of time become accepted.

Keep in mind that this problem-solving system is not only effective in the field, but that it also works in an office environment. While the main thrust of this book is to deal with troubleshooting plumbing problems in the field, this chapter will also provide help for plumbers who own their own businesses and have to deal with routine business problems.

Evaluating the effects of your problem

Once you know you have a problem, you must evaluate its effects before you attempt to correct it. Evaluating the effects of a problem could be as simple as seeing an aging report and realizing your accounts receivables are delinquent and resulting in cash-flow problems for your business.

What is the effect of a dripping faucet? The constant drip can stain and damage the sink it is dripping into, and the cost of wasted water can be considerable over time. There is also the issue of environmental conservation. Both of these examples represent problems and their effects, but the answers are not always so easy to see.

Identifying the problem

Systematic troubleshooting is the way to pinpoint a problem. In theory, troubleshooting is simple; however, in practice, many people (even professional troubleshooters) have difficulty systematically locating root causes of problems. As you implement systematic troubleshooting techniques, you are attempting to identify a problem. Identifying problems correctly can be a complex procedure, but an effective troubleshooting system allows you to tag your problems quickly and correctly.

Problems that are not defined properly cannot be solved positively. For example, if you are looking for what caused water to run through a light fixture in a ceiling of a dining room, it is not enough to know that there is a leak. You must know what type of leak you are looking for—a pressure leak or a drainage leak—and you must find the location of the leak before you can fix it.

If you assume the leak must have come from a bad wax ring under the upstairs toilet, you may be right, or you may be wrong. The leak could be associated with the drainage system of the bathtub. Replacing the wax ring will not do any good if the root cause of the leak is in the bathtub drainage piping.

Developing potential solutions to the problem

After you have identified your problem successfully, you may have to go on a fact-finding mission. Let's assume you have identified your problem as being related to the waste and overflow from the bathtub. You know what the problem is, what the effects of the problem are, and where the problem lies, but you are not sure what to do about solving the problem; this calls for research. Having accurate data to assess is crucial to competent decision making.

Once problems are identified and you have gathered sufficient resource data, your job turns to that of developing potential solutions. Let's assume you have, after weighing the possibilities, decided that the problem is a leak where the tub drain meets the tub shoe. You suspect the root cause to be a lack of sufficient putty under the rim of the drain. At this point you must compile, compare, and eliminate the information you have researched to reach potential solutions. In simpler terms, you no longer have to plan on replacing the waste and overflow. All that should be required is the removal and reinstallation of the tub drain.

Suppose you couldn't find the leak, even though you felt sure it was coming from the vicinity of the bathtub. What would you do? There are times when stepping away from the problem for a new perspective is the most logical move to make.

If you dwell on the same problem too long, you can become frustrated and unproductive. The answer could be right in front of you and go unnoticed at these times. A new perspective can result in immediate solutions to plaguing problems. By giving yourself a short break from thinking about what could be causing the leak, the answer may come to you.

Pulling from past experiences to solve your problem is often an excellent way to find a rapid remedy. Experience is a wondrous thing, and it can give you power—power to be used in effective troubleshooting.

Sometimes the only way to solve a problem seems to be turning to others for help. There is no shame in this; the shame would be living with a problem you were unable to solve because you were ashamed to share it with others. You can turn to coworkers, supervisors, plumbing inspectors, or your employer for help.

In business settings, turning to others for help is common practice. It is often called a task-force approach to problem solving. The more people you have working together to solve a problem, the more likely you are to come up with viable solutions.

Assessing sensible solutions

After compiling a list of potential problem solvers, you must assess the list for sensible solutions. In the case of the leaking bathtub, your list of problem solvers might include the options of replacing the tub waste or removing and reinstalling the tub drain. This is a vital element in problem solving.

When you want to solve a problem well, rather than simply solving it temporarily, the assessment of possible solutions is critical. There are often multiple solutions to any given problem. Finding the best solution requires a system, a system like the one you are being shown.

Picking a plan of attack and implementing it

When you know what you want to do to eliminate a problem, you must pick a plan of attack. In working with the bathtub, your plan of attack is to remove the existing drain, install a ring of putty under the rim, dope up the threads, and reinstall it. With the plan of attack defined, you are ready to proceed with the implementation of your plan.

Tracking your results

You must monitor the results of your problem-solving decisions. In a business setting, this might mean keeping tabs on where your calls from customers are coming from. Are you getting your work from your advertisement in the phone directory, or are most of the calls coming in from your direct-mail campaign?

In the case of the leaking bathtub, tracking the results of your decision is as simple as testing the tub drain several times to be sure the leak has been stopped.

Making progressional adjustments

Tracking your results allows you to make progressional adjustments that make your solutions the best they can be. As this pertains to the tub, if the leak still exists, you can tighten the drain a little more. In the case of advertising, if the direct-mail campaign is your most effective form of advertising, you can budget more money for direct mail and less for other forms of advertising. To guarantee success in problem solving, you must track your results.

Now that you have a broad understanding of what an effective troubleshooting system consists of, let's see how you will use the various modules of the system to create an efficient problem-solving system for your needs.

Three Ways to Make Problems Go Away by Ignoring Them

There are three ways that I have found to make problems go away by ignoring them. This section of the chapter applies more to routine business problems than it does to field work, but some of it applies equally well to both sides of the plumbing trade.

Put time on your side

The first way to make problems go away by themselves is to put time on your side. By this, I mean to simply let a problem run its course and solve itself. This is rarely appropriate in the field, but it can pay off in the office.

What are some of these types of problems? Problems with employee morale could fall into this classification. It can be a serious and profit-threatening problem. If your employees are not motivated to do their jobs properly, your company is going to suffer.

When you are faced with only one employee who is feeling down, normally the problem will go away if ignored. People lose their drive for numerous reasons—problems at home, problems with coworkers, and problems with their supervisors. If you have a good employee who becomes unconcerned about job performance all of a sudden, the problem will probably leave as quickly as it came.

Should the problem persist for more than a day or two, evaluate the circumstances leading up to the lack of performance. Employees are people, and people can be difficult to understand. The smallest event can trigger a lack of interest in an employee.

When I lived in Northern Virginia, I could predict the production my company was going to get from employees by which football team won on Sunday. If the Washington Redskins lost, employee morale and production was going to be down. However, when the Redskins were winning, workers were in good moods and generally quite productive.

The outcome of a football game shouldn't affect a person's work habits, but it can. Counseling employees who are suffering from depression when favorite football teams lose is ludicrous. If you wait a few days, the right teams will win and your employees will be back to normal. The win-loss statistics of professional sports is only one of the insignificant events that can turn employees into slow-moving robots.

The weather affects the attitudes of a number of people. It is natural for a cool, cloudy day to make employees want to stay in bed a little longer. You can't change the weather, and it is doubtful you can change the personality swings in your employees who are affected by the weather. About all you can do is wait for a sunny day.

If a customer has run the well that supplies drinking water to his or her home dry, there isn't much a plumber can do about it. Given enough time, the problem will probably solve itself as the well replenishes its water supply.

Stepping away from the problem

Stepping away from the problem to find a new perspective is frequently an excellent approach to solving complex problems. This is, in effect, a technique that requires ignoring the problem. While the problems this procedure is used on may not go away on their own, you may find a solution faster by ignoring the problem.

There have been many times during my plumbing career when I became so intent on solving a problem in the field that I could not do anything right. I rarely have leaks in my solder joints, but I've had occasions when no matter how hard I tried, I couldn't keep a joint from leaking. When I fought with the problem, it always seemed to get worse. I found that by leaving the problem for a while and cooling myself down, when I went back to solder the joint, the solder took well and the leak was fixed.

Have you ever tried to think of someone's name that you knew but just couldn't remember? Did you think hard for a long time without remembering the name? This is a common problem that many people have. Have you ever noticed how when you quit trying to think of the name and move on to another activity, the name pops into your memory? You may not know it, but you solved the problem by ignoring it.

Delegating a task force

Another way to ignore a problem successfully is by delegating a task force. This is easier for administrative people to do than it is for the plumber in the field. When a task force is put into action, the problem is not being ignored, but *you* are ignoring it. Once you have empowered the task force, you can move on to other responsibilities.

If you assign a task force to bring you potential solutions, your time is better managed, and the people delivering possible alternatives are likely to produce ideas you would not have thought of. You can then review their findings and render a positive decision.

Ten-Step Troubleshooting System

I have a 10-step troubleshooting system that I have used successfully over the years. The principles have been used both in the management of my business and in the troubleshooting of field problems. Let me share this fantastic system with you.

Blinded by what appears to be obvious

The first step in my system requires you to avoid being blinded by what appears to be obvious. Many people believe they know the answer to a problem when they see what appears to be the obvious cause. It is seldom that the root cause of a problem is obvious. Solving part of a problem or the wrong problem is not effective.

Consider this example. You are on a job where the floor is wet around a toilet. This is normally a sign that the wax seal has become ineffective and must be replaced. If you act on the obvious, you will pull the toilet and replace the seal. Often this will be the right move to make, but it may not be.

I was on a job like this just last week. The customer told me the toilet was leaking. The floor all around the base of the toilet was wet, and I suspected the wax seal. However, rather than pulling the toilet immediately, I performed some troubleshooting techniques.

The first step I took was to flush the toilet and watch for water seeping out from under the base. I didn't see any. This perplexed me. At this point, I did a close inspection of the tank-to-bowl gasket and the tank-to-bowl bolts. I found no evidence of water.

The toilet refilled itself, and I flushed it again. The ballcock nut wasn't leaking, no water was seeping under the base, and there appeared to be no reason for the water being on the floor. Then I noticed something, something that should not have been happening.

A stream of water was running down the right side of the toilet tank. The water was coming out from under the tank lid and running down the tank, dripping quickly onto the floor. My new assumption was that for some reason the tank was overflowing.

I removed the tank lid and immediately came face to face with the problem. The flexible refill tube had separated from the clip that should have been holding it in the refill tube. Water pressure was forcing the tube up and causing it to spray water at the junction of the tank and the lid. The water was then running down the tank and flooding the floor.

I reinserted the tube on its clip and taped it into place to make sure it stayed there. Then I toweled up the water from around the toilet and flushed the water closet several more times. The problem was solved.

If I had acted on the obvious, I would have gone through a lot of work that wasn't necessary and still not have solved the problem. Can you see how you might be blinded by the obvious? Don't let this happen to you. Always confirm the root cause of a problem before you attempt to fix it.

Use a systematic approach to identify the problem

You must use a systematic approach to identify the problem properly. If you don't use a controlled system to troubleshoot your problem, you may go around in circles and never expose the best solution. A random process of elimination is not the best way to troubleshoot plumbing problems.

Never assume anything.   Never assume anything about your problem or solutions. The only good solutions are proven solutions, and the only good problems are solved problems.

Create a problem description and profile.   To solve a problem positively, you must have an accurate description and profile of it. You must know what the problem is and what it isn't.

Use a process of elimination.   The process of elimination is an effective method for problem solving when it is used properly. However, it can be a means of wasting time and postponing the satisfactory termination of a problem. The process of elimination should be used, but it should be used systematically on components proven to be likely causes of the problem.

Log your findings.   Log your findings as you progress through the troubleshooting process. Written notes will be invaluable during the solution-assessment segment of your problem solving.

Beware of fringe interference.   Beware of fringe interference that appears to be a cause of the problem. Don't confuse effects with causes. For example, suppose your truck won't start, and you have promised a customer that you will be on the job in 30 minutes. What is the problem you must solve? The fact that the truck won't start is a problem, but it is not the root cause of your most important problem. The primary problem is the fact that you are unable to keep your scheduled appointment with your customer.

The root problem with the truck not starting is in some component of the truck. It may be a dead battery or a failed starter, but you must not allow yourself to focus on the effect (the truck won't start); you must move systematically through your options. These options may involve calling another plumber to cover the service call that you will be late for and then calling a tow truck to have the truck transported to a professional mechanic. If you know something about automotive mechanics, you may contemplate rolling up your sleeves and checking the components that might be keeping the truck from starting.

The first problem you have to solve is that of the customer who is expecting you. The second problem is getting the truck back up and running. If you focus too tightly on the truck, without thinking of the customer, you might wind up with neither problem solved.

Ask questions.   Ask plenty of questions. Ask yourself questions, and talk with your customers. Ask them for background information that might help you to diagnosis the plumbing problem they are having.

Many plumbers don't ask their customers questions pertaining to the history of a problem. This is a mistake. Your customers have more knowledge about the history of their plumbing system than you do. They know if a sink has always drained slowly or if their shower has never had adequate water pressure. Take the time to ask questions, and by all means, pay attention to the answers you receive.

List possible causes.   You will develop ideas for possible causes of your problem as you are troubleshooting it. List these possible causes for evaluation once you have gathered all of the available data. By writing down your suspicions as you go along, you won't forget any of them, and you will be better able to find a good solution to the problem.

Evaluate possible causes and develop potential solutions

When you have completed the troubleshooting process, you are ready to evaluate possible causes and develop potential solutions. This is done by reviewing the notes of your findings and first impressions of possible causes. Once you have developed a list of potential solutions, you are ready to move into the final phase of your troubleshooting and problem-solving system.

Six Methods for Finding Sensible Solutions to Problems

There are six methods for finding sensible solutions to your problems that I believe are very helpful. Let's take a look at them on a one-by-one basis.

Establish your goal

Before you can come to a sensible solution, you must know what you want the outcome of your solution to be. The desired outcome is your goal. In the case of the leaky bathtub that we discussed earlier, your goal is to stop the leak.

Perform a risk assessment

No decision is without risk. To arrive at an acceptable solution, you must establish risk parameters. To illustrate this, let's say that the tub drain is still leaking after you have removed and reinstalled it. Should you tighten the drain more? If you do, you might pop the porcelain on the tub. If you don't, you've still got a leak to contend with.

Most experienced plumbers have a good feel for how much pressure they can exert on a tub drain without hurting the finish of the fixture.

As a calculated risk, you would probably tighten the drain a little more. This is a form of risk assessment where if you're right, the problem is solved. If you're wrong, and the porcelain pops, you made your situation worse than it was to begin with.

Compare competing solutions.

Most problems have multiple potential solutions. These potential solutions will be of a competitive nature, and you must compare them closely to define the best solution. For example, let's look again at the leaking bathtub drain.

When you perform your risk assessment, you have two options: tighten the drain or replace it. If you can tighten the drain successfully, that is the best solution. When the risk is too great to justify making another turn on the threads, you might find that replacing the drain is a better alternative. This is how you compare competing solutions to find the best one.

Play the "what-if" game

You can eliminate competitive solutions by playing the what-if game. Take each possible alternative and ask yourself questions about how the various options will affect your problem and your goal. Some solutions create new problems, others are merely quick-fix techniques, and then there are the best solutions: the solutions that provide lasting satisfaction.

In the case of the leaking tub, cracking the finish on the tub would create a new problem. Putting some type of quick-sealant compound on the outside of the drain would be an attempt at a quick-fix solution that would probably fail. Only stopping the leak with proper plumbing techniques, and without harming the tub, would be the best solution.

Let the best solutions simmer for a day or two

Most problems around the office can be placed on a shelf for a day or two. Field problems don't normally fall into this category, but administrative duties can. Once you have developed your prime prospects for solving the problem, put them on a back burner. Let them simmer and cure. Come back to the solutions with a fresh attitude and review them. The winning selection will probably jump out at you.

Make a decision and move on

Once you have gotten to the decision-making stage of your problem solving, make a decision and move on. Don't linger and agonize over your decision. Make the best decision you can from the data you've gathered and move on. If you study your decision too long, you will become confused and will fail to make a decision. Make the decision and monitor it.

Track the results of your solution and make progressional adjustments as needed. If you must abandon the first decision, based on its performance, don't become frustrated. Go back to the list of potential solutions and review them for a new approach. The important thing is to remain positive at all times.

Develop Your Own System

You can use the suggestions I have given you here to develop your own system for troubleshooting and problem solving. I've found the above steps to be very effective in both my administrative and my field duties. I imagine that you will modify the system, but that's fine. Find a system that works for you and use it.

Why should you develop your own troubleshooting and problem-solving system? Because it will save you time and money. It will also help to keep your customers happy. To expand on this, let's turn to the next chapter.

2

How Effective Troubleshooting Saves Time and Money

This chapter is going to show you how effective troubleshooting saves time and money for you and your customers. In addition to saving time and money, good troubleshooting skills will save you a lot of frustration and some occasional embarrassment.

Some experienced plumbers feel they know it all and have done it all. If there is a plumber who does know it all and has done it all, I'd like to meet that individual. I've been in the trade for about 20 years, and I still learn new techniques and gain new field experience regularly. There always seems to be something that pops up that I've never seen before.

Regardless of how much experience you have, there will always be opportunities to expand your knowledge. The day you feel you know it all and have done it all, you're in trouble. What makes me so wise on this issue? Well, it's not mystical powers, it's experience.

When I was in my mid-20s, I thought I was the best plumber around. I was pretty good, but not as good as I thought I was. Over the last 10 years or so, I've learned a lot more than I knew when I thought I knew it all. It was not until I started to get older and gained more life experience that I was able to reflect on my past and see how stupid some of the notions I had were. You will be doing yourself a big favor if you can avoid falling into the know-it-all trap.

If you've been working in the field for a few years, you no doubt have developed some troubleshooting skills. Right now you may feel that you don't need to alter the way you look for problems in plumbing systems. If you feel this way, I'm willing to bet your attitude will have changed by the time you finish this book. I believe you will find that the recommendations in this chapter, and in the following ones, hit home and prove themselves very valuable.

If you keep an open mind, I think you will find numerous ways to improve your troubleshooting skills. When you do this, you become a more efficient and a more profitable plumber.

If you presently work for an employer and don't think saving time will help you make more money, you might be right. Plumbers who are in business for themselves can cash in on the time they save, but plumbers who are employees don't see the same financial rewards. However, if you become an outstanding troubleshooter, your employer will see that your value to the company has increased. This could result in a raise in pay or, at the least, the chance to keep your job while others are losing theirs.

There are always a few people who are not receptive to new ideas. These people have their own way of doing things, and they have no desire or intention to change their old habits. This group of people limit their professional advancement severely. To prosper professionally, you must be willing to learn on a continual basis, accept change, and roll with the punches.

Much of the information in this book is based on jobs I've done in the past. Not only are you going to learn a dynamite way to find plumbing problems faster, you are going to gain the benefit of my 20 years of mistakes. I may not be able to tell you how to do everything right, but I can certainly tell you a lot of things to avoid doing.

In this chapter we are going to concentrate on the benefits of becoming a super troubleshooter. You learned the basic system that I use for troubleshooting both office and field problems in the last chapter. Here you are going to see why you should bother to put my suggestions to work for you. In the chapters that follow, you are going to get extensive details that are related to troubleshooting specific aspects of plumbing systems. By the time you're done reading this book, you should possess the knowledge to become one of the best troubleshooters in the field. Now, I'd like to go ahead with our discussion on how effective troubleshooting can save you time and money.

Time Is Money

Everyone has heard the old saying, time is money. There is a lot of truth in those words, especially for plumbers who are doing contract work. If a plumber is working as a piece worker or under a fixed-rate contract, lost time is lost money. At the hourly rates most plumbing companies are charging today, it doesn't take much wasted time to amount to a substantial financial loss.

Even plumbers who bill their time on an hourly basis for service work stand to lose money if they are not competent troubleshooters. This money can be lost in more than one way. For example, an incompetent troubleshooter who must charge the customer for time spent finding a problem that a solid troubleshooter would have found more quickly will probably not see much return business from the customer. Once the customer catches on, the incompetent plumber's phone won't ring.

The best service and repair plumbers in the business occasionally have to handle nonpaying callbacks. These annoying and profit-puncturing calls don't come in frequently for conscientious plumbers, but they do raise their ugly heads from time to time. Any plumbing contractor will tell you that callbacks are killers. Having to do work a second time, and without getting paid for it, is no way to run a business.

A lot of callbacks can be avoided by doing the job right the first time and checking the work closely before leaving the job. Even so, sooner or later, you're going to pick up the phone and find that you have to respond to one of the dreaded callbacks. When this happens, time is of the essence.

There are two big problems with callbacks. First of all, customers who have paid to have a job done and who are still experiencing problems, even after paying professional plumber's fees, are usually a little upset, sometimes very upset. The quickest and best way to calm these disgruntled customers is to respond quickly and to solve their problems with a minimum loss of time.

For the plumbing company, callbacks mean negative cash flow. Time is being spent on a callback that could be being charged out at a handsome rate on a paying proposition. But since callbacks must be done free of charge, the company is losing money. If you happen to own the company, this is particularly disturbing.

Honing your troubleshooting skills to a sharp edge will help you cut back on the financial losses caused by callbacks. If you can get in and out of the job quickly, and successfully, you will have appeased the customer and minimized your lost revenue. Plumbers who have to go back to jobs and stumble around looking for the cause of the problem that generated the callback are losing their credibility in the eyes of their customers and some substantial earnings from their day.

Contract Work

Contract work is where a lack of solid troubleshooting skills can cost a plumbing company the most money. Jobs that are put out to bids for contract approval usually don't offer a lot of profit margin. The margin is sometimes so thin that a few hours of unexpected work will turn a good profit percentage into a dismal one.

You may be thinking that contract work is rarely affected by the troubleshooting skills a plumber possesses. If you are, think again. You are partially correct; most contract work is straightforward. It doesn't often involve hunting down problems before they can be fixed. However, there are many times on contract jobs when troubleshooting ability is needed. Let me explain with some examples.

I'm going to give you four examples of how the profit from contract work can ultimately depend on your proficiency with troubleshooting chores. You may have had problems similar to the ones you are going to read about happen to you in the field. I know from firsthand experience that predicaments like the ones about to be described can occur. I've had them happen to me.

A new house

Plumbing a new house seems simple enough. It is all new, clean work, and what could possibly go wrong that would require a plumber to know anything more than the basics of troubleshooting? It wouldn't seem that there would be much call for sophisticated problem solving in plumbing a house, but there can be.

I once plumbed a new house and was finishing the job as it got dark outside. Even though night was close at hand, I wanted to test out the piping and get it ready for inspection. In those days, I used to test with water. After putting full pressure on the copper water lines, I found a leak. It wasn't a big one, but it was enough to make me drain the system and fix it.

I drained the pipes and soldered the fitting again. The solder seemed to take well, and I was satisfied the leak was fixed. Once the joint cooled, my helper filled the pipes with water again. The joint still leaked. Frustrated and hungry, I decided to cut the bad fitting out and replace it.

We drained the pipes and replaced the defective joint. The pipes were holding some water, but it didn't seem like enough to be a problem. I cranked up the torch and soldered the new fitting. The solder ran around the joint like it should, and I assumed that the job would be done in just a few more minutes.

When we charged the pipes with water pressure, I heard that nasty sound of air rushing past my solder joint. Water soon followed. I was starting to lose my temper. My helper wanted to come back the next morning and fix the problem, but I was committed to seeing that leak plugged.

As I recall, it took several more attempts at soldering the joint before I realized that it just wasn't going to take. This problem occurred very early in my plumbing career, so I was not well prepared with alternative methods for soldering wet pipes. During one of my soldering attempts, I noticed the solder was bubbling. The leak was in the exact spot where the bubbling had taken place. I figured out that steam was blowing out of the fitting on the back side and creating the void that was causing the leak.

By the time I had figured out what was going wrong, I was so angry that I could hardly think straight, let alone solder a good joint. Fighting the desire to destroy the piping, I went outside and stared up at the stars that had come out in the night sky. As I stood there, the answer came to me.

I went to my truck and got a bleed coupling. When I went back into the house, my helper was sitting on a bucket and wondering whether I was going to take an ax to the pipe or try to fix it again.

Going back up the ladder, I cut the pipe and put the bleed coupling near the fitting that was giving me so much trouble. With the cap and seal removed from the coupling, steam could vent through the hole as I soldered the joints. As the flame from the torch embraced the pipe and fitting, steam and hot water was expelled through the drain hole in the coupling.

When we pumped the pipes back up with water pressure, the fitting held. The leak was fixed, I was victorious, and I had learned a valuable lesson. If I had known what to do the first time the fitting leaked, I could have saved myself about an hour and half of time, a lot of frustration, and the wages paid to my helper. Plus, I wouldn't have been starving to death by the time I got to eat supper.

When all of this happened, my time was worth $30 an hour. My helper was charged out at $20 an hour. That means my lack of experience and troubleshooting ability cost me $75 in billable time and a lot of personal aggravation.

The remodeling fiasco

This second story is about a remodeling fiasco that cost me several hours of billable time and nearly cost me one of my best contractor accounts. By the time this job took place, I had seasoned into what I thought was a top-notch plumber. My reputation in remodeling work was excellent, and I enjoyed all the work I wanted, at good wages.

On this particular remodeling job I had installed a new toilet in a converted attic bathroom. The contractor I was working for had used me for all of her plumbing for quite awhile, and her company kept a plumber and a helper busy 1 or 2 days a week. This was such a good account that I usually did most of the work myself.

Well, I installed the toilet and drove the 25 miles back to my house. I hadn't been home 2 hours when the phone rang. It seemed the toilet was leaking. This was strange because I had tested the toilet many times before leaving it. Anyway, I got back in truck and drove all the way back to the job.

When I arrived, I flushed the toilet, and sure enough, it was leaking around the base. I pulled the toilet, inspected the wax ring, and reset the bowl with a new ring. The old ring was deformed. It looked as if I had pinched it with the base of the toilet when it was set. The test on the new installation went well, and I left.

Two days later I got another phone call. The toilet was leaking again. I responded to call quickly and found that the toilet was leaking, but I had an excuse this time. The tile contractor had been to the job and pulled and reset the toilet after installing the new tile floor.

Late that same afternoon, my pager went off. It was my favorite contractor calling to tell me that the toilet was leaking again. This was getting embarrassing, not to mention expensive. I went back to house and checked the toilet. It was leaking.

Fed up, I pulled the toilet and put a regular wax ring on the flange. Then I stacked another ring, one with a horn rim on it, on top of the first ring. I reasoned that the tile installation had raised the bowl to where one ring wasn't quite cutting it. After bolting the bowl back to the flange, I tested it a bunch of times. No problems.

For days afterward I waited for the phone to ring with another complaint on the bad-boy toilet. It didn't happen. The double ring worked. Looking back on that job, I remember the flange being low to begin with. The carpenters had built up the floor with underlayment, and when the tile was installed the flange was low. I should have realized then that the flange had to be raised or that double rings were needed, but I didn't.

I've never figured out why the toilet would test okay and then start leaking later, but it did. However, if I had known then what I know now, I could have saved myself a lot of driving, time, lost money, and credibility with both the customer and my contractor friend. I can't remember now how many hours I lost on that job, but there were a lot.

My first garbage disposer installation

My first garbage disposer installation on galvanized drain piping was an eye-opening learning experience. When this job took place, I was in business for myself and working as a contract plumber for a major, national chain of stores. The store sold customers plumbing appliances and provided plumbers, like me, to install them under the store's name and warranty. The money I made as a piece worker for the store was not great, but it was steady, and it helped to keep the bills paid.

The store called me one day and scheduled the installation of a garbage disposer. All of the work done for the store was done on a flat-rate basis. Some jobs I made out on, and some jobs I didn't do so well on, but it evened out and made the working arrangement worthwhile in the early stage of my business development.

As I recall, the fixed-rate fee the store offered me for the garbage disposer installation was $45. Now, that's not bad money for 15 years ago, so long as things go pretty much as they should. This job, however, didn't go anywhere near the way I envisioned that it would.

I picked up the disposer and rolled on the call. All in all, the actual installation went well. It was what happened afterward that was a disaster. The day after the disposer was installed, I got a call from the store. The drain from the kitchen sink, where I had installed the disposer, was stopped up. Since the pipe hadn't been stopped up before my work, I was being held responsible for the complaint.

Not wanting to lose my large corporate account, I responded to the clogged drain. I sent a $\frac{3}{8}$-inch spring-head snake down the drain and cleared it. It took about an hour, not counting driving time.

A few days later, the store called again. The people for whom I had installed the disposer were very upset. Their kitchen drain was stopped up again. I responded and opened the drain again.

The third time I was asked to snake the drain free of charge, I got a little touchy about the subject. This time I went under the house and inspected the drains. That is when I discovered that the kitchen drain was run with galvanized pipe and short-turn fittings. I put 2 and 2 together and figured out that the waste from the disposer was getting clogged up in what was probably already a messy drain line.

I cut out a section of the drain and confirmed my suspicion. The customer refused to pay to have the piping replaced. When I called the store and informed them of what I had found, they refused to pay me to replace the piping. I put the pipe back together with a rubber coupling and left the job. The pipe was draining, but I doubt if it stayed clear for long.

That was the end of my relationship with the department store, but the beginning of my education on galvanized pipe and garbage disposers. Since that day, I have special wording in my contracts for disposal installations on galvanized pipe that protects me from such money-losing situations.

A sewer ejector installation

This story is about a sewer ejector installation one of my employees performed some years ago. It is not a job I'm proud of, but it is a good example of how troubleshooting skills come into play on contract work.

My company was hired to install a basement bath in a townhouse. Part of the installation required the installation of a sewer ejector sump and pump. A plumber who worked for me did the job. He was middle-aged and had a lot of experience under his belt. When the job was done, my plumber tested the fixtures, but he didn't test the sump. The plumbing inspector looked the job over and didn't find any problems. It didn't take the homeowner too long, however, to get on the phone to me. His brand-new fixtures were backing up; the pump wasn't pumping.

My first thought was a blown circuit breaker, but I sent the plumber who installed the job to check out the problem. He called me some time later and said that he couldn't figure out what was wrong. The pump was running, but the fixtures were not draining.

Unhappy, I left home and headed for the job. When I got there, I took one look at the installation and shook my head. My plumber could tell that I was disgusted. When he asked what I was upset about, I pointed to the check valve; it was installed backward. There was no way the pump could evacuate the contents of the sump.

My plumber had been looking for the problem for over an hour, and I found it in less than 5 minutes. Fortunately, it was not a big job to turn the check valve around.

As you probably know, most check valves have arrows on them that show the direction of flow. How could a plumber install a check valve backward in the first place? Secondly, how could the inspector and the plumber fail to see that the check valve was installed backward? Well, for all the possible reasons, money was lost, and a customer was upset.

There you have it

Well, there you have it, four prime examples of how money can be lost on contract jobs when you or your plumbers don't have strong troubleshooting skills. These examples are only a very small percentage of the types of problems you may be called upon to solve in order to salvage profits from contract jobs.

Now that you've seen some proof that troubleshooting is a necessary part of every plumber's job description, let's look at some examples of various jobs where knowing what to look for, where to look for it, and when to ignore certain possibilities can save you time and money.

Forget about Them

There are times when you are troubleshooting that the best thing to do with some types of solutions is to forget about them. If you have enough experience and a structured troubleshooting procedure, you can rule out some causes and cures very quickly. To demonstrate this fact, let's look at the example in the following paragraphs.

Assume that you have responded to a service call where the homeowner is complaining that the toilet in the upstairs bathroom is flushing slowly. The fixture is not stopped up completely, but it has a lazy flush. It is not uncommon for some of the bowl's contents to remain after the toilet is flushed. What will you do, and in what order will you do it?

It would not be unusual to think that the problem is in the drainage piping. In fact, a lot of plumbers would come to this conclusion and look no further until they were sure the drain was not stopped up. This could result in a considerable amount of lost time, the use of a large snake, and facing the homeowner with egg on your face.

There are a number of reasons why a toilet might not flush properly; you will find most, if not all, of them in Chap. 3. In the scenario I've given you here, the first logical step to take would be the removal of the tank top. If some water conservationist has lowered the volume of water in the tank too drastically, that could be the problem. This can be checked quickly and easily.

Assuming the water in the tank is at the proper level, you must look a little further, but don't pull the toilet just yet. With the tank lid off, flush the toilet and see how the bowl reacts. It's possible that the flapper or ball that covers the flush valve might be coming down too soon, restricting the water available to the bowl for a full flush.

If the troublesome toilet passes both of these tests, inspect the flush holes. These are the little holes that are hidden under the rim of the toilet. Flush the fixture to see if water streams down the sides of the bowl. If it doesn't, probe the flush holes with a piece of stiff wire, like a coat hanger.

Houses that get their water from wells often experience problems with mineral deposits forming in the flush holes. A few firm jabs with some stiff wire will tell you if calcium or some other obstruction is blocking the holes.

Assuming everything has checked out okay so far, try putting a closet auger through the bowl. If it brings back wax, you can assume that the wax ring has spread out over the drain and is impeding the flushing action.

If none of the earlier suggestions prove fruitful, it would be wise to inspect the piping under the toilet. Long vertical drops into short-turn fittings can make a toilet flush poorly.

Wall-hung toilets can suffer from slow flushing if the fitting on the carrier is a short-turn fitting. You'll find more about this in the next chapter.

With all the easy options ruled out, you can resort to snaking the drain. If your snake doesn't get the job done, read Chap. 3 from start to finish.

The type of progressional troubleshooting steps I've shown you here are exactly what you need to practice in the field. When you have a systematic approach to problem-solving, the task always works out better.

What possible solutions did this job offer that you could forget about? Every single step you took gave you one more option that you could rule out. As you worked your way through the job systematically, you were able to forget about problems that may have plagued other plumbers. There is also another way to forget about possible solutions and move on to more productive possibilities.

Some jobs present symptoms that rule out certain causes immediately. For example, the toilet job we just looked at could not have been associated with the ballcock, unless there wasn't enough water in the tank. Once you knew there was adequate water in the flush tank, you were able to rule out low water volume and the ballcock.

As a rule, a lazy flush doesn't have anything to do with incoming water pressure. This eliminates possibilities of problems with the supply tube and cutoff valve, so long as the tank has adequate water in it for a good flush. See how easy this stuff is?

Move Methodically

If you want to become an expert troubleshooter, you must learn to move methodically. Chasing after false causes and fixing things that are not broken are sure ways of losing both time and money. You must discipline yourself to face problems with an attitude that will allow you to win.

Patience is a quality that many young people have trouble grasping. Most young plumbers don't want to take the time to think a problem through before taking action. They want to jump right in and start working on something, even if it is the wrong something.

With maturity and experience, plumbers learn the value of assessing problems carefully before acting. I used to be quick on the trigger when it came to fixing a problem. As soon as I knew something was wrong, I would dig in and start making adjustments, repairs, replacements, or anything else it took to get results. Unfortunately, not all of my results were good ones. Things have changed for me; I've learned the value of patience and observation skills.

I talked with my parents not long ago and related my present-day procedures to them. In the conversation, I explained how that in the old days I would look once and act one, two, three times, or more if necessary, to fix a problem. Nowadays, I look three times and act once. It has taken me a long time to develop my patience, but the results have been worth the wait. There is certainly something to be said for advanced planning.

If you learn to take your time, make keen observations, plan your work, and work your plan, you will enjoy an easier, more enjoyable, and more profitable career. It really is amazing that there is truth to the old saying about working smarter instead of harder.

I'm nearly 38 years old. My physical condition is not nearly as good as it was when I was 23 and the best plumber this side of the Rockies, at least in my own eyes. Even so, I find that I'm a very effective plumber. Younger plumbers can move better and faster than I can, but when the day is done, I've usually accomplished more, and I've done it with less wear and tear on my body and mind.

Discipline is an important factor in troubleshooting. When you are faced with a difficult problem, your nature may gnaw at you to make a rash decision. Having the confidence to hold back and work through the problem thoroughly before acting is not always easy, but it is usually effective.

Having a good plan, and sticking to it, is the key to success as a troubleshooter and problem solver. If you take pot shots blindly in the dark to find the reasons why some plumbing fixture or device is not doing what it is supposed to, you will waste a lot of time and energy. A pinpointed plan of attack will almost always produce better results.

Knowing What to Look For

Knowing what to look for is paramount to your victory as a troubleshooter. If you don't have enough experience or knowledge, there is no way you can solve problems quickly and effectively. Much of what you learn will come the hard way, from in-the-field trials and tribulations. You are going to make bad calls from time to time, but don't worry about that; learn from it.

My mother used to tell me that the first time people do something the wrong way is a learning experience. The second time they do the same thing improperly is a mistake. If you follow this philosophy, you can face your troubles with a good attitude and a winning approach. Don't look at your bum calls as failures or mistakes; look at them as experience. If you find yourself making the same poor choices repeatedly, you should start to fuss at yourself.

No troubleshooter is perfect, but a lot of them are very good. Most great problem solvers have achieved their status from hands-on experience and relentless reading. You are reading at this very moment, and that indicates to me that you have the potential to be a super troubleshooter. It is not that my words will give you the wisdom to conquer the plumbing world, but I'm sure you will learn aspects of the trade from me that you have not experienced personally.

Even if you only pick up a few helpful tips, you're money ahead. Gaining knowledge is a lot like building a snowman. When you start out to build a snowman, you have empty hands. You cup some snow in your gloves and form it into a ball. Then the ball is rolled through the snow, and it gets bigger and bigger, one snowflake at a time. Before you know it, you have a sphere of snow so big it can't be lifted.

Learning to troubleshoot plumbing is not so different from building a snowman. You start with the knowledge you already have. Then you read books, talk to other plumbers, get some on-the-job experience, and before you know it, you're rolling all that information up inside your brain. Just as the little snowball grew into a massive sphere, your knowledge of the trade will balloon.

Every day that you work in the field is a learning experience. Whenever you share war stories with colleagues in the supply house, you are picking up pointers that can help you out. Reading books like this one is an ideal way to learn from the experiences of others. Once you roll all of this knowledge into one tight ball of memory, you are in a much better position to solve problems.

Problem Solving

Problem solving is a skill, much the same as plumbing is a trade. You may view yourself as a plumber, but to be a great plumber, you must be much more. To be a successful plumbing contractor, you have to extend your reach into even more areas.

It is rarely enough to be just a good plumber. If you're working for an employer, they are going to expect more from you than just soldering pipes and drilling holes. They will want you to keep track of your truck inventory, your time sheets, and a whole lot more. They will also expect you to think for yourself.

I've worked as a supervisor in shops where the plumbers in the field practically refused to think for themselves. When something, anything really, was not the way it should be, they would call the service manager for advice. This type of activity might boost the service manager's ego, but it does nothing to raise the level of appreciation an employer has for the field plumber.

It is understood that rookies have to be trained, and they have to make some mistakes to learn. This is fine. However, when you have mature plumbers, who have been plumbing all their working lives, an employer has the right to assume they know how to deal with some adversity in the field.

I had a plumber working for me many years ago who should have known a lot more than I did. He may have known more, but if he did, he never showed it. This guy was an employer's nightmare. Some days he was good, and other days he was as bad as it gets. I kept him because he seemed to try, but eventually, the problems he caused outweighed the benefits I gained, and I had to let him go.

The biggest problem that I've seen with plumbers who were not good troubleshooters was their inability to think ahead. Gifted troubleshooters can plan several steps ahead of where they are, and they can envision what is likely to happen before it happens. Let me tell you a quick story that is bound to bring a smile to your face.

I had a plumber once who had good intentions but only modest skills. His limitations were known to me, so I could work around his weak points, most of the time. When I added a new truck to the company, I assigned the job of installing shelves in the truck to this plumber. It was a slow day, and I didn't want to have to send him home without pay. With the best of intentions, the plumber pulled the van into the lot and set to work.

I glanced out the window occasionally and saw that he was working hard, he was even using a level to get the shelves put in properly. My wife and I discussed how well he seemed to be doing, and how we wished he was a little better in the field.

Sometime after lunch the plumber came to my door and asked me to inspect his work. My wife and I went out to the van and looked at the new shelving. The job looked great. We were in the process of congratulating our plumber when he wanted to show us some of the finer details of his work.

The plumber took a 2-foot level and placed it on the top shelf. He was pleased that the bubble read dead level. At first, I was also pleased, but soon it dawned on me that something was wrong. You see, the level said the shelf was level, but the van was parked on an upgrade. When I realized what had happened to my well-intentioned plumber, I didn't know what to say. It was a sure bet he would figure it out sooner or later, so I mentioned the fact that the van was parked on a slight hill. His face paled, and his eyes almost seemed to water.

Without a word, the plumber slammed the side door and backed the van back onto level ground. Kimberley and I could see the van moving as he stomped through it. The side door opened slowly, and the plumber, with his head hanging low, stepped out. His terrific job had turned into a major flop.

I probably should have been angry that he had wasted so much material in building crooked shelves. He was expecting to be ridiculed for his actions, but I couldn't bring myself to yell at him. Instead, I laughed. In fact, I laughed and laughed and laughed. My amusement may have hurt him more than the sharpest words I could have spoken, but my exuberance was not meant to hurt him. I found the situation genuinely entertaining. Kimberley started to laugh with me, and then, finally, the plumber even began to laugh.

Once we were all over the initial reaction to the shelves, my plumber apologized and offered to pay for the new building materials to replace the shelves. He even offered to build them on his own time. With that type of heart and dedication, I couldn't discipline him. We laughed off the whole experience and reworked the shelves on the next rainy day, at my expense.

I've told you this story not because it pertains to troubleshooting, but because it says a lot about many plumbers. My plumber knew better than to build the shelves the way he did. He had to know the truck was on an angle and that the level would not work in the manner he wanted it to. But, he did his job and was proud of it, until he realized his blunder.

Lots of plumbers make similar mistakes in the field. They jump to a conclusion without thinking about existing circumstances or repercussions. If you will take the time to think through what you are about to do, and what the likely outcome will be, you will make fewer mistakes and have more satisfied customers.

The Bottom Line

The bottom line in business is net profit. Effective troubleshooting can increase the net profit for plumbers working on contract jobs and piece work. It also enables hourly plumbers who are faced with callbacks to keep their financial losses to a minimum.

Now that we have learned the structure of good troubleshooting and seen how it can save any plumber time and money, let's get into the specifics of troubleshooting various types of plumbing fixtures and devices. We will start with toilets and work our way through the book until all normal plumbing fixtures and devices have been covered.

3

Troubleshooting Toilets

Toilets are common plumbing fixtures, and they are not particularly complicated, but troubleshooting them can be a frustrating job. They are so simple, in some ways, that when there is a problem, it can be difficult to find. Plumbers with years of service experience have seen most everything a toilet can offer in terms of complications, but a lot of plumbers don't think about flush holes, condensation, and similar hidden defects that can drive a troubleshooter crazy. During my 20-year career, I've seen some mighty fine plumbers baffled by troublesome toilets, myself included.

Troubleshooting is largely a matter of logic, common sense, and systematic deduction of probable causes. As simple as the methodical act of troubleshooting is, it still gives a lot of plumbers plenty of misery. This chapter is going to help you avoid the hours of aggravation I have endured over the years.

In theory, there should be only a limited number of problems that may contribute to a toilet's malfunction. Theory is a wonderful thing, but it does not always prove itself out. I have found toilets with crumpled beer cans in their traps, and I've seen good plumbers replace toilets only to discover that the problem they were faced with wasn't the toilet at all. If you're lucky, you will never come face to face with a problem you can't solve, but the odds are good that if you stay in the trade long enough, you will get stumped by a stubborn toilet problem.

As we move through this chapter, you will learn aspects of toilet trouble that you may never have encountered. The benefit of learning what to anticipate, with the help of this book, will reduce your exposure to embarrassment and possible financial losses on the job. To stress this point, let me give you two real-life examples of how good, experienced plumbers were left dumbfounded by toilets. As we go through these examples, see if you can solve the problem before I give you the answer.

The Commercial Toilet That Couldn't Flush

This first example is about the commercial toilet that couldn't flush properly. The fixture was a wall-hung unit (Figs. 3.1 and 3.2), and it was installed beside another wall-hung toilet that flushed correctly. Both toilets shared the same drain and vent.

A call came in that one of the toilets was not flushing properly. I dispatched an experienced service plumber to investigate the problem. Shortly after arriving on the job, my plumber called in to say the drain was partially plugged and would need to be snaked out. We were busy that day, so I directed the on-site plumber, who had not been able to clear the stoppage with his small drain cleaner, to anoth-

PLACIDUS
Siphon Jet Action
Wall Hung Closet

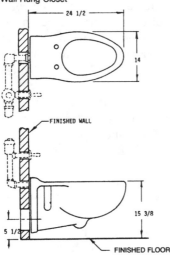

Placidus 3-449

Bowl: 3-449 *Placidus* water economy direct fed siphon jet, elongated rim, whirlpool action, 1 ½" back spud wall hung bowl.

Valve:

Seat:

Support:

Note: Operation – For efficient operation of the bowl, a minimum flowing water pressure of 15 P.S.I. is required at the valve.

Figure 3.1 Wall-hung water closet with concealed flush valve. (*Courtesy of Crane Plumbing.*)

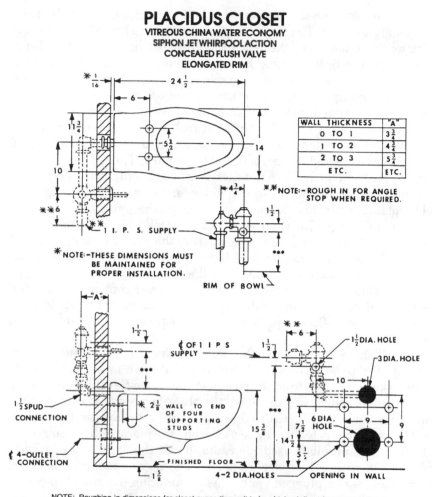

PLACIDUS CLOSET
VITREOUS CHINA WATER ECONOMY
SIPHON JET WHIRPOOL ACTION
CONCEALED FLUSH VALVE
ELONGATED RIM

WALL THICKNESS	"A"
0 TO 1	$3\frac{3}{4}$
1 TO 2	$4\frac{3}{4}$
2 TO 3	$5\frac{3}{4}$
ETC.	ETC.

**NOTE:- ROUGH IN FOR ANGLE STOP WHEN REQUIRED.

1 I. P. S. SUPPLY

*NOTE:-THESE DIMENSIONS MUST BE MAINTAINED FOR PROPER INSTALLATION.

RIM OF BOWL

¢ OF 1 I P S SUPPLY

$1\frac{1}{2}$ DIA. HOLE

3 DIA. HOLE

$1\frac{1}{2}$ SPUD CONNECTION

6 DIA. HOLE

WALL TO END OF FOUR SUPPORTING STUDS

¢ 4-OUTLET CONNECTION

FINISHED FLOOR

4-2 DIA. HOLES   OPENING IN WALL

NOTE: Roughing-in dimensions for closet supporting unit to be obtained direct from manufacturer. Check with manufacturer for proper outlet coupling extension beyond finished wall.

***Variable-check with manufacturer of flush valve for proper roughing-in dimension.

Figure 3.2 Cut-away and rough-in details for a wall-hung water closet with concealed flush valve. (*Courtesy of Crane Plumbing.*)

er call and told him to inform the property owner that a drain-cleaning crew would be there soon.

When the next plumber arrived with the big drain-cleaning machine, he took the initiative to investigate the problem prior to snaking the drain. In less than 5 minutes he found that the first plumber was mistaken in his diagnosis of the problem. How did he come to this conclusion so quickly?

It was simple logic that led him to the first elimination in the troubleshooting process. The first plumber had been shown the defective toilet by an employee of the retail building. Acting only on the designated toilet, the first plumber had never checked the neighboring toilet. When my second mechanic arrived, he flushed both toilets, and the one on the upstream side of the problem toilet flushed correctly. This indicated that there couldn't be any obstruction in the primary drain pipe or vent. In the short time it took to flush the second toilet, the troubleshooter had ruled out snaking the main lines. This saved the plumber time and the property owner money. There was, however, still the problem with the one toilet that would only swirl and drain very slowly when flushed. This called for further troubleshooting.

The plumber ran his closet auger through the toilet bowl and met with no resistance. He then removed the toilet from its bracket to inspect the drain outlet. Using a mirror, the plumber inspected all parts of the toilet's trap and flushing mechanisms; he found no problems. He then turned his attention to the drain arm.

With nothing being visibly wrong, the plumber connected a garden hose to a silcock and ran a full stream of water into the pipe's drain outlet. The water flowed perfectly, proving their was no stoppage in the line.

With all these steps taken, the plumber was convinced the problem was with the toilet fixture, and that it would have to be replaced. He obtained authorization from the customer to replace the toilet and did so.

Once the new toilet was installed, the plumber tested his new installation. It was then that he realized he had made a mistake in his diagnosis. The new toilet acted the same as the one the plumber had replaced.

Baffled, the plumber used his radio to call the shop and discussed the problem with the service manager. The service manager believed the problem was a plugged vent. This assumption put the plumber on the roof with a snake to clear the stoppage in the vent. Some time later the plumber was back on the radio with the service manager. Snaking the vent had done no good in solving the problem.

The service manager came to me and asked for my opinion. He filled me in on all the steps that had been taken to resolve the problem. After listening to what had been done, I had some questions of my own. You are about to find out what the real cause of the problem was. Have you figured it out yet?

I asked the service manager how long the toilet had been flushing slowly. He didn't know. I suggested that the plumber check the fitting at the drain outlet to be sure that it was an approved, long-turn fitting and not a sanitary tee. The service manager gave me a funny look and left my office shaking his head.

Well, it turned out that I solved the problem without ever leaving my office and without ever seeing the toilet. After talking with the store owner, my plumber discovered that the toilet had never worked properly, but that the owner had just put up with it until he had called us.

Upon further inspection, the plumber found my guess of a sanitary tee being used improperly to be correct. He cut out the fitting and replaced it with a long-turn fitting. When the toilet was reinstalled, it worked like a charm.

In this case, two experienced plumbers and a seasoned service manager had overlooked a very simple explanation. The first plumber didn't check enough options before making a judgment call. The second plumber did a good job of troubleshooting, but his lack of experience prevented him from thinking about the carrier fitting. As for the service manager, he should have figured the problem out but didn't.

How was I able to solve the problem so quickly and easily? I used troubleshooting information supplied by the second plumber to eliminate possible causes. When I thought about all that had been done, the carrier fitting was the only option left to consider.

If I had been the plumber on the job, I would have followed basically the same steps the second plumber did, but I would have inspected the carrier fitting when flushing the line with the water hose. This would have saved time and the purchase of an unneeded new toilet.

If you determined the cause of the problem on your own, congratulations. If you didn't, you now have one more valuable piece of experience in your head to help you if you run across a similar situation. Now, let's look at another example and see how well you do in guessing the cause of the problem.

A Residential Toilet

This true story is about a residential toilet (Figs. 3.3 and 3.4) that would not flush with enough force on a consistent basis. Sometimes the toilet would evacuate its contents, and sometimes it wouldn't.

The plumber who responded to the call for service had years of experience in residential service plumbing. His skills were good, and his dedication to perfection was superb. After checking the toilet, he found that it often swirled nearly to the flood rim before gradually subsiding. The first thing the plumber checked was the water level in the flush tank; it was okay. Then he ran a closet auger through the trap and met with no resistance. Next he checked the other plumbing fixtures in the house. They all drained properly, ruling out a full septic tank or clogged sewer. It was then that he removed the toilet from its flange and did a visual inspection of the toilet bowl and the drain opening under the closet flange. He found no problems evident.

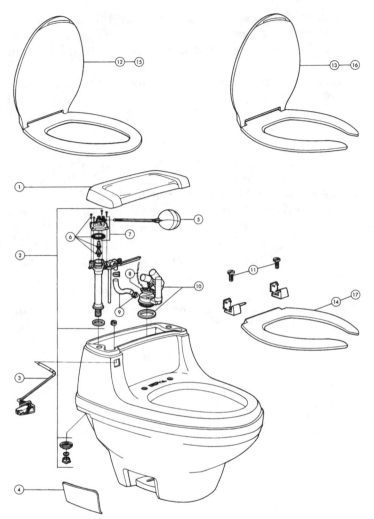

Figure 3.3 Detail of parts of a flush-tank toilet. (*Courtesy of Delta Faucet Company.*)

Confused, the plumber called me. I asked if the wax seal was intact or if it may have been obstructing the drain in the toilet bowl. The wax was fine (Fig. 3.5). I asked if he had checked the flush holes for mineral deposits. He hadn't, and that seemed a logical next step. The plumber hung up, and I went back to my office work.

A little later I heard the plumber talking to the service manager on the radio. The homeowner had authorized a new toilet, and my plumber was on his way to get one. Later in the day, my phone rang again. This time my plumber was calling to explain that he had

MODELS — A

KEY	PART DESCRIPTION	FINISH	ORDER NUMBER	4000	4000 OF	4000 OFC	4001	4002	4007	4100	4100 OF	4100 OFC	4100 OF VRL	4100 OFC VRL	4101	4102	4106	4107	4108	4109
1.	Tank Lid	White	RP8083	●	●	●				●	●	●	●	●						
		Bone	RP8084				●													
		Parchment	RP8085					●												
		Antique Red	RP8574														●			
		French Vanilla	RP9791								●									
		Cerulean Blue	RP12072																●	
		Americana Brown	RP12073																	●
2.	Ballcock Diverter Assy. (Complete)		RP8076	●	●	●	●	●	●	●	●	●	●	●	●	●	●	●	●	●
3.	Trip Lever Assy. (Complete)	Chrome	RP9469	●	●	●	●	●	●	●	●	●	●	●	●	●	●	●	●	●
4.	Mounting Bolt Covers (2)	White	RP8087	●	●	●				●	●	●	●	●						
		Bone	RP8088				●													
		Parchment	RP8089					●												
		Antique Red	RP8575														●			
		French Vanilla	RP9792								●									
		Cerulean Blue	RP12074																●	
		Americana Brown	RP12075																	●
5.	Arm & Float		RP8077	●	●	●	●	●	●	●	●	●	●	●	●	●	●	●	●	●
6.	Bonnet Screw, Diaphram, Plunger Assy. & Valve Seat		RP8124	●	●	●	●	●	●	●	●	●	●	●	●	●	●	●	●	●
7.	Complete Bonnet Assy. & Diaphram		RP8125	●	●	●	●	●	●	●	●	●	●	●	●	●	●	●	●	●
8.	Flapper & Chain		RP8091	●	●	●	●	●	●	●	●	●	●	●	●	●	●	●	●	●
9.	Hose & Clamps		RP8090	●	●	●	●	●	●	●	●	●	●	●	●	●	●	●	●	●
10.	Flush Valve Assy. (Complete)		RP8092	●	●	●	●	●	●	●	●	●	●	●	●	●	●	●	●	●
11.	Seat Bolts (2)		RP8075	●	●	●	●	●	●	●	●	●	●	●	●	●	●	●	●	●
12.	Seat-Regular*	White	RP8072	●																
		Bone	RP8073				●													
		Parchment	RP8074					●												
		French Vanilla	RP9498						●											
13.	Seat-Regular OFC w/Cover*	White	RP8164		●															
14.	Seat-Regular OF without Cover*	White	RP8435	●																
15.	Seat-Elongated*	White	RP9489								●									
		Bone	RP9490														●			
		Parchment	RP9491													●				
		Antique Red	RP9492														●			
		French Vanilla	RP9493															●		
		Cerulean Blue	RP12070																●	
		Americana Brown	RP12071																	●
16.	Seat-Elongated OFC w/Cover	White	RP9799												●	●				
17.	Seat-Elongated OF without Cover	White	RP9800											●	●					

Figure 3.4 Description of parts shown in Fig. 3.3. (*Courtesy of Delta Faucet Company.*)

installed a new toilet, but that the problem persisted. I agreed to go to the job and see what I could do.

When I arrived, the plumber showed me the old toilet and the problem he was experiencing with the new toilet. He took me into the basement to show me that he had run a snake through the cleanout under the toilet with no success. As soon as I entered the basement, I knew what the problem was. Have you got any idea at this point what was causing the trouble?

My experienced plumber had been standing right in front of the problem and had never noticed it. The vertical riser from the building drain came out of a sanitary tee and extended upward for about 4 feet. Now have you figured it out?

The vertical drop was so great that when the toilet was flushed, the water would hit the back of the tee and splash back up the riser. This

Figure 3.5 Standard installation of closet bolts and wax ring on a closet flange.

back pressure would not allow the toilet bowl to empty properly, and toilet tissue was being left in the riser because the water was dropping too quickly. The vertical riser from the building drain to the closet flange was much taller than the plumbing code allows, and it was the primary cause of the problem.

We cut out the illegal piping and installed a wye and an eighth-bend to shorten the vertical drop and to give a smoother transition from the toilet to the building drain. While my plumber didn't argue with me on this decision, neither did he think my repair idea was going to work; he was wrong. After the new piping was in place, we tested the toilet over and over again. It flushed perfectly every time.

This was another case where a new toilet was bought unnecessarily. Fortunately, the homeowner was not angry, and she wanted to keep the new toilet. In both this case and the previous one, I reduced the labor charges to reflect what the customers would have been charged if the plumbers had been better troubleshooters. As a business owner, I don't like to give money away, but there was no way I could charge those people for my plumbers' mistakes.

What is the moral of these two stories? It is that knowing proper troubleshooting techniques will save you time, money, and embarrassment. Both of the plumbers in the above examples were experienced journeyman plumbers, but they couldn't find the solution to their on-the-job problem. Now that you've seen how easy it is for experienced plumbers to overlook obvious causes of problems, let's delve into the troubleshooting procedures for toilets in a step-by-step, problem-by-problem manner.

Flush-Tank Toilets That Run

Toilets that run constantly are not very difficult to troubleshoot. There are a number of possible causes for this type of problem, but they are easy to identify and eliminate. Let's look at each of the possibilities individually.

The ballcock

The ballcock is a likely suspect when it comes to a toilet that won't stop filling itself with water. There are two basic types of ballcocks. One style uses a horizontal float rod and a float ball. The other type has its float on the vertical shaft of the ballcock. Both types work on the same principle. When the volume of water in the flush tank is sufficient to raise the float of a ballcock to a predetermined point, the valve closes, and the water is cut off. This is a very simple principle, and one that is easy to check. Let's look first at the type of ballcock that uses a float rod.

Float-rod ballcocks. Float-rod ballcocks are the most common type found in toilets. The ballcock rises vertically in the flush tank, and the float rod extends from the top of the valve into the flush tank. One end of the rod is attached to the ballcock, and the other end has a float ball screwed onto it. The ball may be made of styrofoam, or it may be a hollow, plastic ball.

If you are inspecting the refill tube from the ballcock and find water running through it, you can assume your running toilet is the victim of a faulty ballcock. It is, however, possible that the ballcock is leaking water into the tank from some other point. To test for this, you should first cut off the water supply to the toilet.

With the water cut off, flush the toilet. Once the tank is drained, hold the float of the ballcock in an extreme upward position and turn the water supply to the toilet back on. No water should come from the ballcock. If the ballcock produces water under these conditions, it would be a good idea to replace it.

While most ballcocks can be repaired, they are so inexpensive that you are usually doing yourself, and your customer, a favor to replace them. If you replace a diaphragm and leave the job, you may be called back to correct a similar problem with the old ballcock. This aggravates customers, and it is common to have to do the work without charge under a warranty basis. Unpaid callbacks are bad for your boss or your business.

If the ballcock works properly in the dry test, release the float and observe the filling action of the ballcock. Does the float ball rub against the side of the flush tank? If it does, you need to bend the float rod to allow the ball freedom from obstructions.

After the tank is filled, does water run over the overflow tube? If it does, the float rod needs to be bent downward. This will cause the valve to cut off sooner, resulting in the solution to your problem.

Should you have a persistent problem in adjusting the float rod, check the float ball. Float balls sometimes become waterlogged (filled with water). The weight of the retained water can be enough to cause the ballcock to malfunction. All that is required to correct this situation is the simple replacement of the float ball.

Vertical ballcocks.   Vertical ballcocks (Fig. 3.6) have their floats attached to the vertical shaft of the valve. There is no float rod to contend with, and the float is not likely to take on excess water.

You can check a vertical ballcock with the same methods described earlier. If the float needs to be adjusted, there is normally a small chrome device that allows you to control the position of the float. By squeezing the small piece of metal between your thumb and

Figure 3.6 Vertical ballcock.

forefinger, you can slide it up and down to adjust the position of the float.

Many of these types of ballcocks are designed to be adjusted for different heights in flush tanks. This requires turning the entire shaft, and the ballcock must be removed from the tank to accomplish the task. It is unlikely that you will need to go to this extreme but be aware that the option exists.

If the nature of the problem is a toilet that runs continuously, the solution most likely lies in the ballcock. If the problem is a ballcock that cycles on and off at various intervals, the problem is more likely to be in one of the other elements of the toilet.

The overflow tube

The overflow tube in a flush tank can cause the toilet's water supply to run endlessly. Many inexperienced plumbers overlook ways that are possible for this to occur. It is not uncommon for service plumbers to look only at the top of the overflow tube. They assume that if the water level is not above the edge of the tube, the tube couldn't be the problem. This is not always true.

Many new toilets have plastic overflow tubes, but older toilets are equipped with brass tubes. These tubes are screwed into the flush valve and offer two distinct possibilities for hard-to-find leaks.

The thin threads of a metal overflow tube can disintegrate over time. If a hole develops in the tube or around the threads, water can leak into the toilet bowl. Since this leak is below the water surface in the flush tank, it can be difficult to find.

The plumbing in houses served by well water is frequently affected by minerals and acid found in the water. It is possible for acidic water to eat holes in any portion of a metal overflow tube. Water seeping into these holes and running into the toilet bowl can create a problem that may seem to be the ballcock's fault.

With either of the above situations, a ballcock will have to cut on periodically to replace the water lost through the overflow tube. If you believe the overflow tube is leaking, replace it.

Flush valves

Many new flush valves are made from plastic materials, but there are still a great number of metal flush valves in use. Just as overflow tubes can be affected by various water conditions, so can flush valves.

Metal flush valves can become pitted with depressions or holes. These pitted areas create a void between the flush valve's surface and the tank ball or flapper that seals it. A strong magnifying glass and a good flashlight make these conditions easier to find.

To check for a pitted flush valve, you must first cut off the water supply to the toilet and drain the flush tank. If you find pitted areas, you have two choices. You can use sandpaper to smooth out the pitted surface, or you can replace the entire flush valve assembly. Sanding the surface area will normally provide some temporary success, but the best way to ensure long-term satisfaction is replacement, preferably with a plastic flush valve.

Tank ball or flapper

The tank ball or flapper can be the cause of a running toilet. As time passes, these devices can become worn to a point where they no longer seal the opening in the flush valve. If this happens, water will leak into the toilet bowl, and the ballcock with be forced to cut on at periodic intervals. Inspect the seating surface of the tank ball or flapper and replace it if the rubber looks worn or damaged.

The lift rods and guide for a tank ball can also contribute to a running toilet. If these items work their way out of adjustment, the tank ball cannot seat properly. It is not uncommon for the lift-rod guide to become misplaced. If this happens, you can reset its adjustment, and the problem should be resolved.

In the case of flappers, the tank handle or chain can prevent the device from seating properly. By flushing the toilet several times and observing the movement of the flapper, you will be able to detect any problem with its alignment.

If the tank handle rubs the side of the flush tank, it may be holding the flapper up, allowing water to run through the flush valve when it shouldn't be. A chain that is too short or tangled can have the same affect on the flapper's ability to seal the flush opening.

Commercial Flush-Valve Toilets That Run

Commercial flush-valve toilets (Figs 3.7 and 3.8) that run require you to disassemble and inspect the flushing mechanism. Since the toilets don't use flush tanks, they don't have ballcocks, flappers, ank balls, lift wires, refill tubes, overflow tubes, or the tank-type flush valves found in flush-tank toilets.

To troubleshoot these toilets, you must be familiar with the inner workings of commercial-grade flush valves. Because of the washers, diaphragms, and numerous other possible culprits in defective flush valves of this nature, it is wise to install a replacement kit of all the essential elements to solve your problem. These kits are readily available for all major brands of flush valves, and they are the fastest, surest way of remedying the defect. By replacing all the small parts, you leave little chance of a recurring problem.

Toilets That Flush Slowly

Toilets that flush slowly can be devilish to troubleshoot. You saw in the two examples I gave early in the chapter how even experienced plumbers can be fooled by slow-moving toilets. There are, to be sure, many possible causes for a lazy flush. Let's take some time now to explore these possibilities and identify ways to find and correct the causes of lackluster flushes.

Commercial flush valves

Commercial flush valves can be responsible for slow flushes. If the valves don't release an adequate amount of water, the toilet bowl cannot be cleansed properly.

If a commercial flush valve is not producing the desired amount of water, check the control stop. Remove the chrome cover plate and use a screwdriver to make sure the valve is open to its proper position.

Assuming the valve is open and the flush valve is still not working properly, you will have to inspect the inner workings for defects or obstructions. Check the water-saver device to make sure the disks are working properly. If you find no obvious obstruction in the working

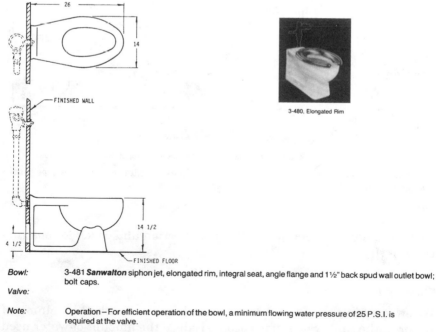

3-480, Elongated Rim

Bowl: 3-481 **Sanwalton** siphon jet, elongated rim, integral seat, angle flange and 1½" back spud wall outlet bowl;
         bolt caps.

Valve:

Note: Operation – For efficient operation of the bowl, a minimum flowing water pressure of 25 P.S.I. is
         required at the valve.

Figure 3.7 Floor-mounted flush-valve-type water closet with concealed flush valve. (*Courtesy of Crane Plumbing.*)

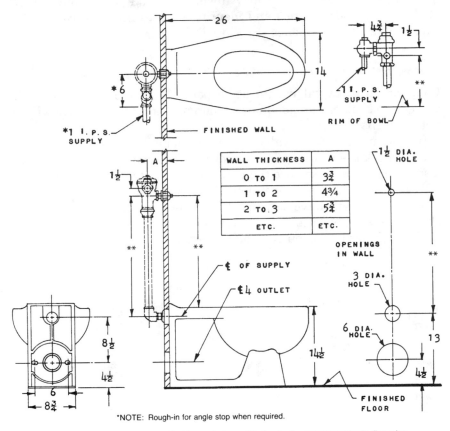

SANWALTON
VITREOUS CHINA
SIPHON JET
CONCEALED FLUSH VALVE
ELONGATED RIM, INTEGRAL SEAT
WALL OUTLET BOWL

WALL THICKNESS	A
0 TO 1	$3\frac{3}{4}$
1 TO 2	$4\frac{3}{4}$
2 TO 3	$5\frac{3}{4}$
ETC.	ETC.

*NOTE: Rough-in for angle stop when required.

**Variable-check with manufacturer of flush valve for proper roughing-in dimension.

Figure 3.8 Cut-away and rough-in details of a floor-mounted flush-valve-type water closet with concealed flush valve. (*Courtesy of Crane Plumbing.*)

parts of the valve, remove the flush connection and check for obstructions in the spud hole.

Flush tank

When working on a toilet with a flush tank, check the volume of water in the tank. Sometimes people reduce the amount of water used in a flush tank to conserve water and save money; this can result in a slow flush. If the water level is low, adjust the ballcock to bring the water up to the fill line. When there are water-saving devices placed in the tank (such as bricks), remove them and make sure the tank fills to the fill line.

Once you have a full tank of water, flush the toilet. It is a good idea to put toilet tissue in the bowl before flushing. This simulates an actual flush better than an empty bowl would. If the paper is not removed quickly and entirely from the bowl, you still have some troubleshooting work to do.

Malfunctioning flappers and tank balls

Malfunctioning flappers and tank balls can cause the water supply to a toilet bowl to cut off prematurely. If the flapper or ball drops too quickly, the flush hole will be sealed before enough water has passed to cleanse the toilet bowl. This problem can usually be corrected by adjusting or replacing the flapper or tank ball assembly. It may require repositioning the toilet handle, chain, lift wires, or lift-wire guide.

As an example, a chain that is too long may not pull a flapper far enough up so that it stays suspended long enough to properly flush the bowl. By shortening the chain, the flapper is pulled farther up, and the water rushes out of the tank to flush the bowl. This is a very simple procedure, but it is one many professional plumbers fail to notice.

Flush holes

I have had experienced plumbers on my payroll who didn't know what flush holes were. Flush holes are the small holes around the inner flush rim of a toilet. They can't be seen by looking in a toilet bowl unless a mirror is used or you are able to get your head below the rim and look up.

These holes can become obstructed with mineral deposits and inhibit the normal flushing of a toilet. If you flush a toilet and don't see numerous streams of water running down the inside of the bowl, there is a good chance the flush holes are plugged up.

The flush holes are designed to clean the sides of the bowl, and if they aren't able to function, the toilet may flush slowly. When these holes become blocked, a sturdy piece of wire, such as a coat hanger, is all that is needed to clear them. Pushing the end of the wire into the holes will normally break up the mineral deposits and return the toilet to good working condition. Occasionally, the deposits will be too large to remove, and the bowl will have to be replaced. Problems with flush holes occur most often with toilets served by well water, but they are worth checking under any conditions.

Obstructed traps

Obstructed traps can certainly account for a slow flush from a toilet. Traps can become blocked by any number of foreign objects, some of which a closet auger will pass right by without clearing. Strange objects in traps probably account for most of the mysterious malfunctions of slow-flushing toilets. Since this condition is so prevalent and so often misunderstood, let's take a few moments to look at some of what my plumbers and I have removed from the traps of water closets. The details of what objects we have discovered and how they nearly avoided detection will prepare you for your own future surprises.

A beer can.   On one occasion I found a beer can in the trap of a toilet. When you consider the size of an opening in the bowl of a toilet, it is hard to believe a beer can could find its way inside, but I have seen the evidence, and it can happen. The can had been crumpled and crushed, but it was all in one piece.

The beer can was found early in my plumbing career, and its presence in the trap had me befuddled. I had augered the trap, pulled the toilet, snaked the drain, and all but given up on fixing the fixture. I assumed the toilet must be replaced. I was about to call the owner of the property to get authorization to replace the toilet when I decided to take one more look at the toilet.

With the toilet laying on its side, I looked into the drain from the base of the bowl. I couldn't see any reason for a problem. Then I ran my closet auger through the trap and saw it come out clean. It was not until I went to set the toilet back upright that I noticed anything.

When I picked the toilet up I heard something rattle inside it. I removed the flush tank and shook the bowl. There was clearly something inside it that didn't belong there. I fished around in the trap with a wire coat hanger and could feel something, but I couldn't retrieve it.

While I was working on the toilet the owner of the property came in. I explained to him what I had found and asked how he wanted to proceed. He knew the bowl would have to be replaced, and we were both curious about what was making the noise. The owner asked me to break the bowl open, and I did. That's when we found the beer can.

The property owner assumed that one of his rental tenants must have caused the problem. I installed a new toilet bowl and went on about my business, a little smarter for my experience.

The beer can, in its crushed condition, was acting as a baffle. It would allow water to pass through the trap, but it would catch solids and tissue paper, causing the toilet to drain slowly, if at all. Even the closet auger would slide right past the can without interference. As you can imagine, this type of problem is hard to identify.

A rubber ball.   One of my plumbers was faced with an unusual situation when he encountered a rubber ball in the trap of a toilet. In his case, the closet auger was meeting resistance that it could not get past. Water would drain slowly from the toilet, but solids would cause overflowing conditions. This was another occasion when the cause of the problem was not identified until the toilet bowl was broken open.

Other strange encounters.   Some of the other strange encounters with foreign objects in toilet traps have accounted for a rather unique collection of plumber's experience. I have found all types of toys in toilet traps. Wash rags, hair curlers, shampoo tops, makeup containers, and a host of other items have all served to make toilets flush slowly. Most of these items were never dislodged with a closet auger. They were found either by breaking the bowls open or with a mirror. A long-handled inspection mirror and a flashlight can make your trap troubleshooting much more effective.

Wax obstructions

Wax obstructions can contribute to early aging in plumbers. When a toilet is set, the wax ring sometimes spreads out too far over the drain opening. If the wax protrudes from the flange and hangs out over the drain, it can catch a lot of toilet tissue. In time, this buildup can make emptying the toilet bowl a slow process.

A wax obstruction will usually show up on the end of a closet auger. If you put your auger through the trap and it comes back with wax on the end of it, be prepared to pull the toilet.

When you remove the toilet from the flange, you will probably find a disfigured wax seal. You have to scrape off the old wax, install a new ring, and reset the toilet. If the toilet was not set down on the flange straight and evenly, or if the bowl was twisted during installation, there is a chance for a wax obstruction.

The bad thing about this type of problem is that it doesn't show up right away. The toilet can be tested and work perfectly, only to begin giving trouble a few weeks down the road.

Partially blocked drainage pipes

Partially blocked drainage pipes can cause toilets to drain slowly. If the drain lines have any type of buildup in them, like tree roots or grease, the full flow of the pipe will be restricted. This condition can go unnoticed with other fixtures in the house, but the volume of water produced from a flush may expose the problem.

The small amount of water going down the drain from a sink or bathtub may not be noticeably slow. It may be that the drain only seems to act up when a toilet is flushed. This can be a difficult problem to put your finger on. When everything else in the house seems to be draining correctly, it would lead you to believe the toilet is at fault. There is, however, an easy way to check for such a problem.

When you are faced with a possible drain blockage, remove the toilet from its flange. Take a 5-gallon bucket of water and pour it down the drain at the closet flange. Follow that water with another bucket of water as quickly as possible. If the drain can take 10 gallons of water quickly, the odds are the problem is in the toilet. If, however, the water gurgles and backs up, check the drain and vent system.

Illegal drainage piping

Illegal drainage piping is yet another cause of slow-flushing toilets. If a pipe doesn't have the proper amount of fall on it, the toilet draining into the pipe may flush slowly. As you saw in one of the earlier examples, a long, vertical drop can cause a toilet to malfunction. If you have access to the piping under a toilet, check it out for illegal fittings, improper grade, or excessive riser heights.

Toilets That Overflow

It is easy to pinpoint trouble in toilets that overflow. The problem is either in the toilet's trap or in the drainage piping. A closet auger is a fast way to check out the trap. If the auger goes through smoothly, the problem is probably downstream.

If the contents of the drainage piping are backing up into other fixtures, such as a bathtub, you can automatically rule out the toilet trap; you know you have a blockage in the drainage pipe.

If you remove the toilet from its flange and find standing water in the pipe, you know you are dealing with an obstructed drainage system. With houses that use septic systems, it would be wise to check the septic tank before investing a lot of time and effort in snaking the drain. Homeowners often forget to have their septic tanks pumped, and that can result in a complete backup of sewage in the house.

Another consideration before you begin to snake drains is to inquire about the presence of any type of sewage pump. If the toilet is connected to a sump pit with a pump in it, the pump may be failing. If after checking all of these possibilities you haven't found a reason for the backup, you will have to snake the drain.

Toilets That Condensate

Toilets that condensate can fool plumbers. If the tank is condensating in a hard-to-see area, the water developing slowly on the bathroom floor may appear to be coming from the base of the toilet. This would indicate a faulty wax seal, when in reality, it is only condensation.

Condensation typically occurs when there is an extreme temperature variance between the water in the toilet and the room temperature. This can easily be the case for toilets supplied by well water during a hot summer.

Some homeowners dress their toilets up in cloth covers. These covers can conceal condensation and lead an inexperienced plumber on a wild goose chase. If you have unexplained water on the floor around a toilet, inspect carefully for condensation before removing the toilet from its flange.

There are a few ways to reduce the problems associated with condensation. A liner, usually made of insulating foam, can be installed on the interior of a flush tank to reduce condensation. These foam panels are glued to the tank, so the tank must be completely dry prior to their installation.

A second alternative is to install a condensate tray under the tank of the toilet to catch any water droplets. This arrangement can be ugly, and it will do nothing for a toilet bowl that is condensating.

The third, and most effective, option is to install a mixing valve on the water supply to the toilet. The mixing valve allows hot and cold water to commingle in the toilet tank, eliminating any chance of condensation.

Toilets That Fill Slowly

Toilets that fill slowly are usually in need of a new ballcock. It is sometimes possible to adjust an existing ballcock to increase its flow rate, but replacement is usually the most logical course of action.

A crimped supply tube can cause a toilet to fill slowly. Every now and then a supply tube is hit with a mop or kicked to a point that it crimps. These occasions are rare, but they do happen. Don't neglect to inspect the closet supply when troubleshooting for the reason a toilet is filling slowly.

The cutoff valve under a toilet should always be checked to prove that it is in a full-open position. It is possible that debris has gotten into the stop valve or the supply tube. To check for this, disconnect the ballcock nut and check the water pressure at the top of the closet supply tube. If it is inadequate, remove the supply tube and see what the pressure is like at the valve. If you are still experiencing low pressure, look to the water distribution piping.

If a toilet is supplied with water from a galvanized steel water pipe, there is a good chance the water distribution pipe will need to be replaced. Galvanized pipe tends to rust and clog up with age. To determine if the problem is in the distribution pipe, all you have to do is remove the stop valve and check the pressure at the end of the water pipe. Be careful not to flood the bathroom when conducting this test.

To avoid flooding, have an assistant standing by the main water cutoff to quickly turn the water on and off. You won't need much time to see how much water shoots out of the pipe. If you have to work alone, you can loosen the stop valve until water sprays out around the connection between the valve and pipe. By doing this you can tighten the connection quickly and keep the water spillage to a minimum.

Toilets That Leak around Their Bases

Toilets that leak around their bases are usually suffering from one of three problems: The wax ring needs to be replaced, the bowl is cracked, or the drainage pipe is backing up.

The most common cause of water leaking around the base of a toilet is a faulty wax ring. This condition requires the replacement of the existing wax. When the flange is too far below the finished floor level, it may be necessary to use two wax rings, one stacked on top of the other (Fig. 3.9).

A cracked bowl can allow water to seep out and imitate the symptoms of a bad wax ring. Condensation, a cracked flush tank, loose tank-to-bowl bolts, and faulty flush-hole gaskets can also cause water

Figure 3.9 A double wax-ring seal on a closet flange.

to gather on the floor around a toilet. To check for these conditions, all you have to do is rub all of the areas with toilet tissue. The toilet tissue will get wet if there is a leak, and you will find the problem faster than you could with just your eyesight.

When the drain for a toilet backs up, it sometimes spills out under the toilet. This buildup of water can seep out around the base and appear to be the fault of the wax seal. There is generally some evidence that discloses this type of problem to a trained eye that is watching a toilet flush. There will usually be a bubbling or a slight rise in the water as it rests in the bowl. If you suspect this is your problem, pull the toilet and pour 10 gallons of water down the drain as quickly as possible. If there is a partial blockage, the test should reveal it.

The Key to Troubleshooting

The key to effective troubleshooting is a systematic approach. If you use the information in this chapter, and the ones that follow, to eliminate possible causes one at a time, you will be able to find the root cause of your problem. Once you have identified the problem, correcting it will be much easier.

Now let's move on to Chap. 4 and learn troubleshooting procedures for sinks and lavatories.

4

Troubleshooting Sinks and Lavatories

Have you ever had a lavatory that wouldn't hold water but that didn't show evidence of a leak anywhere? There are times when lavatories can drain themselves without a trace of water being left as evidence of a leak. If you think I'm describing a pop-up plug that is not sealing properly, you're wrong. The type of leak I'm talking about can happen even when a rubber stopper is placed in the drain opening. No matter how tightly the drain opening is sealed, the type of leak being described can happen. What kind of leak could empty a lavatory bowl full of water without so much as a drop of water to disclose it? Well, you'll find the answer a little later in the chapter. In the meantime, think about your past experiences and see if you can come up with a viable answer.

Have you ever been curled up under a sink, looking for a leak you couldn't find? Was the base of the cabinet saturated with water even though there was no evidence of a leak, no matter how often you filled and emptied the sink? This situation can be frustrating and uncomfortable physically. A plumber could test the sink in this scenario time and time again and never find the cause of the problem. Why? Because the problem is not a true plumbing problem. Oh yes, the water is there to prove a leak exists, but the leak is not from a faucet or a drainage fitting. What else could be causing the cabinet to get wet? No, it's not in the water distribution pipes, stop valves, or supply tubes. If you don't already know the answer to the question, you will find it a little later in the chapter.

Are you tired of facing questions without answers? Well, that is very much what troubleshooting is; you have questions but no answers. As a troubleshooter, it is your job to take the information you have and assess it for a suitable solution. If you don't have enough information, you must continue testing with a process of elimination to find answers.

This chapter is all about sinks and lavatories. While there are many types of sinks and lavatories in use, most of them share the same basic plumbing characteristics. There are, of course, some differences in how the various types should be treated during troubleshooting. As you go through this chapter, you will find some advice on specific types of sinks and lavatories. Most of the text, however, applies to any type of sink or lavatory.

We will look at all aspects of troubleshooting sinks and lavatories in the next several pages. Because of the vast number and complexities of faucets, they are not covered in this chapter. Faucets and valves will be explained in Chap. 10.

Kitchen Sinks

We will begin our journey with kitchen sinks since they are the most common type of sink. As you probably know, there are many, many types and styles of kitchen sinks available. Some have single bowls; some have double bowls (Fig. 4.1), and some have specialty bowls. The two most common materials for kitchen sinks to be made from are stainless steel and cast iron. Some sinks are equipped with garbage disposers, some with continuous wastes, and some with individual traps. These are only a few of the possibilities for differences in sinks, but even so, they are all about the same when it comes to troubleshooting. Basic principles apply to all sinks and can be used to solve most problems. Let's look now at some common and some not-so-common problems associated with kitchen sinks.

A wet base cabinet

Our first sample service call involves a wet base cabinet beneath a kitchen sink. Put yourself in the place of the plumber on the service-call example you are about to be given.

A homeowner calls your office and complains that the cabinet under the kitchen sink is soaked with water. You ask all the over-the-phone questions needed to determine the urgency of the call. The homeowner has never seen where the water is coming from; it is only known that something is leaking. You schedule an appointment to correct the problem.

When you arrive at the house and look under the sink, you see water stains in the base cabinet. It is obvious a significant amount of

Figure 4.1 Double-bowl, stainless-steel, kitchen sink. (*Courtesy of Republic sinks by UNR Home Products.*)

water is finding its way into the cabinet. Now it is your job to find out where the water is coming from.

The kitchen sink is a stainless-steel, double-bowl model, equipped with a garbage disposer and an end-outlet continuous waste. There are an ice-maker connection and a dishwasher connection under the sink. The faucet is a single-handle model, with a spray attachment (Figs. 4.2 and 4.3).

After your initial look at the circumstances, you suspect the spray attachment is the problem. Your past experiences have proven that kitchen sprays often leak. As your first troubleshooting step, you turn on the faucet and depress the spray handle. Water comes out of the sprayer, and as you watch under the sink, there is no leak. A bit surprised, you continue your troubleshooting procedure.

You fill both bowls of the sink to capacity and then release the

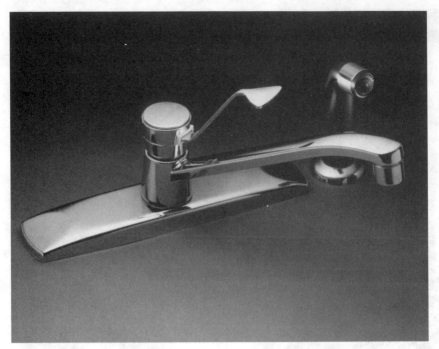

Figure 4.2 Single-handle kitchen faucet with spray. (*Courtesy of Moen, Inc.*)

water. A close inspection of the drainage systems reveals no evidence of a leak. The problem is becoming more difficult to explain. The faucet isn't leaking, the drainage connections are not leaking, and the spray attachment is not leaking. This leads you to a close inspection of the supply tubes, stop valves, and compression connections. After a thorough investigation, these items are not found to be defective.

As a last resort, you fill the side of the sink where the garbage disposer is connected. When the sink is full, you release the water and turn on the disposer. Still no leak. Scratching your head, you are perplexed at what could be causing the leak.

You take a paper towel and rub all the water-supply connections; the towel remains dry. When you apply the same towel test to the drainage connections there is no dampness. Many plumbers would be calling their service managers for advice at this time, but you are determined to solve the problem on your own. The only problem is, you don't know what else to check. What would you do?

Do you know where the water is coming from yet? I'll give you a hint; it has to do with caulking. Now do you know what test to conduct?

The leak in this case is not a true plumbing leak. The water supply

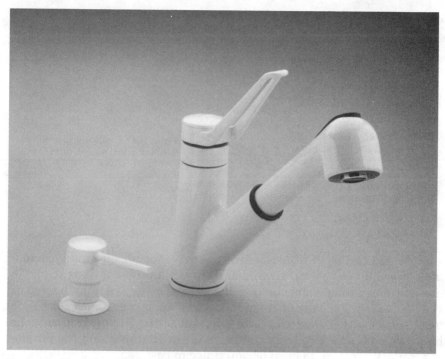

Figure 4.3 Single-handle kitchen faucet with pull-out spray. (*Courtesy of Moen, Inc.*)

and drainage connections are all sound. The water is running from the kitchen counter under the rim of the sink and dripping onto the base cabinet. This leak will not show up during any standard plumbing test. The only way to identify this type of problem is by putting water on the counter, around the edges of the sink.

When homeowners use their kitchen sinks, water frequently gets on the surrounding countertop. If the rim of the sink is not caulked properly, the water will seep under the edge of the sink and drip into the base cabinet (Fig. 4.4). A simple bead of fresh caulking will correct the problem, but many plumbers fail to find the source of the water. If the cause of a problem cannot be found, the problem cannot be fixed.

As a business owner and supervisor, I've run into this type of situation often. Plumbers under my direction have been stumped by such mysterious leaks. When I've asked them if the caulking under the sink was checked, they invariably answer, "No." Once they apply the splash test to the edges of the sink, they are amazed at the volume of water that can flood into the cabinet.

Having this kind of knowledge can mean the difference between

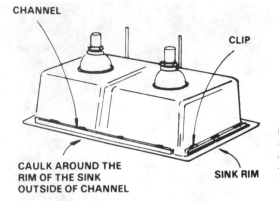

CHANNEL

CLIP

CAULK AROUND THE
RIM OF THE SINK
OUTSIDE OF CHANNEL

SINK RIM

Figure 4.4 Caulking location on a kitchen sink. (*Courtesy of Republic sinks by UNR Home Products.*)

success and failure. Callbacks are expensive for all plumbing companies. If you do not positively find and correct a problem like the one in the above example, someone is going to lose money. It may be you or your boss, but someone is going to have to keep going back to the job until the situation is rectified. In addition to the financial losses, the homeowner is likely to lose confidence in the plumber and the plumbing company. A simple bead of caulking can cost you a customer.

You see, not all troubleshooting in plumbing involves what you may think of as plumbing. There are times when you have to look deeper than the common or obvious causes for leaks.

I once secured a major account from a situation similar to the one just described. A large motel was having problems in their kitchenette units. Many of the units were suffering from wet base cabinets. The owner of the motel had instructed the manager to have the leaks taken care of. The base cabinets were becoming stained from the water, and rot was going to set in if the leaks were not fixed. The manager called in one plumbing company who checked all the plumbing connections and couldn't find a cause for the leaks.

A second plumbing company was called. This company did extensive work in the leaking kitchens. The tubular fittings were all disconnected and reinstalled. New slip-nut washers were installed, new seals were placed under the basket strainers, and the supply-tube connections were all done over. The plumbing company assured the manager that the problems were resolved and would not recur. The cost of the work was extensive, but the manager assumed it was needed work and would have paid the bill, except that the leaks did recur. When the plumbing company came back, they were unable to find out where the water in the base cabinets was coming from. This obviously frustrated the motel manager. That's when she called my company.

We dispatched a service plumber to the motel. When he arrived, the manager filled him in on the last two companies who had done the previous troubleshooting. My plumber was a seasoned veteran, and as soon as he had the background information to go on, he had a good idea of what the problem was.

The manager accompanied the plumber to one of the units where there was water standing under the kitchen sink. The plumber looked under the cabinet and shined his light on the water connections; they were dry. Then he filled the single-bowl sink to capacity and released the water. None of the drainage fittings leaked.

The plumber took a towel out of his work bucket and placed it over the water in the cabinet. Then he spread paper towels out over the heavy towel and asked the manager to watch what was going to happen. The manager was confused, but she paid attention. Reaching into the wall cabinets, the plumber removed a water glass. He filled the glass partially with water and poured the water around the edges of the sink. In a matter of moments, the water was dripping down onto the paper towels. The manager was amazed.

With this problem pinpointed, the plumber and the manager moved on to the other kitchenettes that had leaks. In every case the cause of the leaks was a lack of caulking around the edges of the sinks. About an hour later, the leaks were all fixed. The manager was elated that a few tubes of silicone caulking and less than 2 hours of labor were all that was required to solve her problems. The cost for this work was a fraction of what the other plumbing company had charged, and most importantly, my plumber fixed the problem.

From that day on, the manager never called another plumbing company. The experience that my plumber had was what made it possible to fix the leaks and to secure a great account. I imagine the other plumbing companies could have figured out what the problem was if they had looked far enough, but they didn't. Their concentration on pure plumbing problems left an opportunity for my company to be a hero. By looking beyond the normal reaches of plumbing, my plumber was able to impress the manager in a way that was not forgotten.

A faucet-hole leak

Have you encountered a faucet-hole leak? This type of leak will normally appear to be a problem with the kitchen faucet or spray assembly. Water will have accumulated on the floor of the base cabinet and may be clinging to the supply tubes or spray hose. When you first see the water, it is natural to assume the problem is with the faucet.

In a way, the problem is with the faucet, but it is not a faucet defect or a problem with the faucet connections. The problem is a bad seal between the faucet base and the sink top. It could be a bad gasket or a deteriorated putty seal. In either case, water will leak under the faucet base and run through the faucet-mounting holes in the sink. To correct this problem, the faucet should be removed and a new seal of putty or a new gasket should be installed. Applying a bead of caulking around the perimeter of the faucet base is another option for solving this type of problem.

Basket strainers

Basket strainers can be involved in two types of kitchen-sink problems. If a sink will not hold water, the basket strainer is usually the problem. If the basket has a bad seal on it, the entire basket should be replaced. The existing drain housing can normally remain intact, but the basket insert should be replaced.

A bad seal under the lip of basket-strainer housing can allow water to seep under it and into the base cabinet. If the putty used to create this seal becomes old and brittle, leaks are likely. This type of leak will often seep out to the tailpiece and run down the tubular drainage fittings. The leak may appear to be coming from a slip nut or the rim gasket under the sink drain. To correct the problem, you must remove the basket-strainer housing and install a new seal of putty (Fig. 4.5).

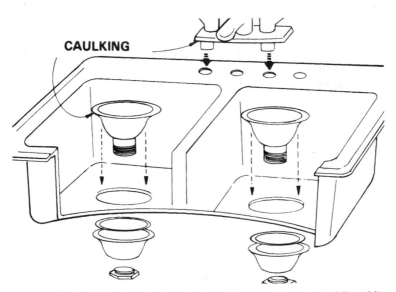

Figure 4.5 Caulking location for basket strainers. (*Courtesy of Republic sinks by UNR Home Products.*)

Garbage disposer drains

Garbage disposer drains are very similar to basket-strainer assemblies in the ways that they may leak. If the rim of the drain collar does not have a good seal of putty under it, water will seep past it. If you find water running down the side of a disposer, dripping off the mounting bracket, or under the disposer, check the putty seal under the drain rim.

Continuous wastes

Continuous wastes and slip nuts account for many of the leaks under a kitchen sink (Fig. 4.6). When people store objects under a sink, they often hit and disturb the tubular drainage pipe and slip nuts. If a slip-nut washer is old and brittle, it doesn't take much stress to make it leak.

Many plumbers make a big mistake when they are looking for leaks in the drainage system of a kitchen sink. I've seen numerous good plumbers make the same mistake when they were testing out new installations and when they were looking for leaks in existing plumbing. They frequently fail to fill the sink to capacity before looking for the leaks.

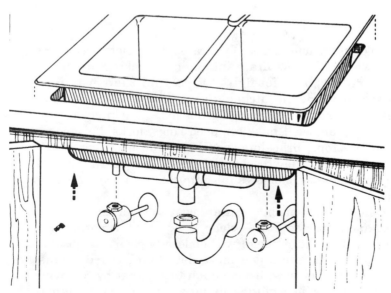

Figure 4.6 Detail of a center-outlet continuous waste for a kitchen sink. (*Courtesy of Republic sinks by UNR Home Products.*)

Slip-nut leaks don't always occur every time water is drained past them. For example, if a kitchen faucet is turned on and the water runs directly from the faucet down the drain, there may not be enough pressure on the faulty slip-nut washer to create a drip. However, if the sink bowl is filled and all of the water is released at one time, the volume and pressure of the water can cause a significant leak at the weak joint.

Homeowners often fill their sinks to capacity to wash dishes. When they release the water, leaks can occur that a plumber will never find without filling the sink with water. This can cause a plumber to tighten the slip nuts and leave the job, satisfied that the leak is fixed. But the next time the homeowner fills the sink to do dishes, the leak will reappear. This creates a callback for the plumber, aggravation for the homeowner, and lost money for the plumbing company.

When you are looking for leaks in the continuous waste of a sink, always fill the sink to capacity and release all of the water at one time. If you are working on a double-bowl sink, fill both bowls and release them simultaneously.

Traps

Finding a leak in a trap should be done with the same procedure that is used on continuous wastes. Fill the sink to capacity and release all of the water at once.

Drainage problems

Kitchen sinks are prime targets for drainage problems. This is especially true when the sinks drain into older plumbing pipes, such as galvanized steel. Many foreign objects find their way into the drainage systems beneath kitchen sinks. Grease, however, is one of the largest contributors to drainage problems in the kitchen.

As a service plumber, I have removed rings, forks, knives, spoons, shampoo-bottle tops, bracelets, earrings, and even aerators from kitchen drains. With the crossbar strainers in modern sink drains, not nearly as many large objects fall into the drainage system as did in earlier years.

Most drainage problems in sinks are easy to troubleshoot. The trap is removed and inspected. If it is clear, the problem is normally in the trap arm or drain, and a snake will generally clear the obstruction. We are going to get into much more detail on clogged drains and vents in Chap. 18. For now, it is enough to say that by removing the trap, and possibly sections of the tubular waste, you can eliminate the possibility of blockages in a kitchen sink.

Bar Sinks

Bar sinks are basically just miniature versions of kitchen sinks. All of the troubleshooting information given for kitchen sinks can be applied to bar sinks.

Laundry Sinks

Laundry sinks don't offer many troubleshooting challenges. These sinks are pretty straightforward, and the problems associated with them are easily found. The same techniques used to troubleshoot kitchen sinks can be applied to laundry sinks.

Laundry tubs that are equipped with pumps, however, are another story altogether. It is not unusual for laundry tubs to be fitted with pumps to lift their contents up to a suspended drain line. These pumps and the objects that find their way into the drains of laundry tubs can create a host of problems. Let me give you a couple of examples from my past experiences.

As you probably know, most laundry-tray pumps have very small discharge lines, usually only about $3/4$ inch in diameter. The pumps themselves are not large, and these two factors combine to create problems. Laundry tubs are meant to be convenient, but when they are set up with pumps, they can be anything but convenient.

My first story involves a laundry pump that one of our plumbers worked on twice before I bailed him out. The plumber responded to a call for a laundry tub with a pump that would not drain. He did his troubleshooting and decided that the check valve was not working properly. This wasn't an unreasonable assumption, but it turned out to be the wrong diagnosis. The plumber replaced the check valve and tested the sink. It worked, so he left.

In less than a week, the customer called back and complained that her sink was once again unable to drain. The same plumber took the callback. He disconnected the drain line and inspected the recently installed check valve. The spring in the check valve was laced with threads of clothing, so he replaced the valve. This replacement lasted about as long as the other one had. The customer called back, in a fury, and fussed and fussed and fussed.

As a damage control effort, I responded to the complaint personally. Before going out to the job, I talked to my plumber and reviewed his previous findings. It made sense that the check valve was failing and allowing the water to run back into the sink, only to keep the pump running constantly.

When I arrived at the house, I inspected the sink and pump arrangement. There was a problem with the check valve. It was jammed open, allowing pumped water to retreat back into the sink. When I investigated further, I found the same lint and clothing threads wound up in the valve that my plumber had found on previous calls. It was obvious that the homeowner was not being conscientious in preventing foreign objects from entering the drain.

Since I had already had to eat one service call and one new check valve, I wanted to make my visit the last one that would be necessary. I questioned the homeowner about the types of things she used the laundry tub for, and she was somewhat evasive. When I asked if she ever used the basket strainer that my plumber had given her and told her to use after the first call, she assured me that she did. Somehow, I doubted that she was telling the truth.

I told the homeowner that I would personally fix her problem. This calmed her, and I went to my truck for parts. One of the parts was a new check valve, and the other was a piece of screen wire. I was sure the woman was not using a strainer in the sink, but at that particular point I couldn't prove it.

I cut a section of the screen wire and installed it between the tailpiece and the drain fitting on the sink. From up above, looking into the drain, the screen was not apparent. When I had replaced the check valve, I instructed the homeowner to always use the basket strainer. In fact, I even had her sign the service ticket that said that the warranty on the work would be voided if she failed to use the basket strainer. She wasn't happy about signing the agreement, but she did.

It hadn't been 3 days since I left the house before the complaining homeowner was shouting in my ear over the telephone. It seemed that her laundry tub was not draining again. As you might imagine, I took the call myself. When I arrived in the home, I questioned the woman about her use of the basket strainer. She assured me she had used it on every occasion. Well, guess what? She lied.

While she was standing in the laundry room, I disconnected the tailpiece and showed her the plugged-up wire that I had placed over the drain. Her eyes got big and her bottom lip started to quiver. She had been caught. I explained that the junk in the wire couldn't have gotten through the basket strainer; I even proved it to her. It wasn't long before she admitted to forgetting to use the strainer, sometimes. I put the sink back together and left the job.

We never got any more work from that customer, but then, I didn't really want her work anymore. Lint and strings can wreak havoc with the check valves in laundry-tray pump lines.

On another occasion, we had a similar problem with a laundry tray on a pump system, except this time, the pump kept seizing up. The plumber who responded to the call got the pump going, but a few weeks later the same trouble was happening all over again. A different plumber took the callback, and he found the problem quickly. What do you think might have caused the pump to jam? Would you believe gravel from an aquarium? Well, that's what it was.

The homeowner in this case washed his aquarium out in the laundry tub. When he did, some of the gravel from the bottom of the fish tank would escape and find its way into the pump. The gravel would lock up the pump and render the laundry tub useless. You can imagine my plumber's surprise when he found colored gravel in the pump.

What is the point to these stories? The point is simple: expect the unexpected. Laundry-tray pumps that attach directly to the fixture are not that great to begin with, and when you factor in negligence, they can become downright troublesome.

Service Sinks

Service sinks are no more difficult to troubleshoot than laundry sinks. While service sinks are often larger and heavier and have different types of traps than laundry sinks, the same troubleshooting principles apply. Most service sinks are equipped with cleanout plugs that allow easy inspection of the drainage system for obstructions.

Lavatories

Lavatories come in many sizes, styles, and shapes. They work on different principles than the other sinks we have discussed. The biggest difference is the presence of a pop-up assembly.

There are a few minor differences in the types of problems you may encounter with various styles of lavatories. For example, a wall-hung unit may have water splashing down behind it, but that is unlikely with a drop-in lavatory. Let's take a quick look at each type of common lavatory and explore problems that may be unique to them. When we have finished this, we will look at the troubleshooting steps for lavatories in general.

Wall-hung lavatory

A wall-hung lavatory should have a waterproof seal installed at the point where the fixture meets the wall it hangs on. If this seal is defective or nonexistent, water can run down behind the fixture and imitate a leak in the drainage or water-supply systems.

It is not uncommon for people to lean on wall-hung sinks; this can break the waterproof seal between the fixture and the wall. If the top of the sink is used to hold soap or if the users of the fixture splash water onto the top of the sink, the water can run over the unprotected back edge and find its way under the sink. If you are looking for the source of a mysterious leak with a wall-hung lavatory, check the seal between the fixture and wall.

Pedestal lavatory

A pedestal lavatory also needs a waterproof seal where its back edge meets the host wall. Since a pedestal lavatory is basically a wall-hung lavatory with a pedestal placed under it, the fixture shares the same characteristics of a wall-hung lavatory.

Drop-in lavatory

A drop-in lavatory should have a waterproof seal installed under its rim and a bead of caulking applied to the outside edges of the rim. Silicone caulking is typically used to create the waterproof seals needed for lavatories.

When a drop-in lavatory is installed without the proper seal between it and the counter, water can run under the fixture. We talked about this same type of condition when we were on the subject of kitchen sinks. Any water that gathers on the counter may be able to slip under the lavatory rim and give the impression of a plumbing leak.

Rimmed lavatory

A rimmed lavatory should have a waterproof seal under the support rim. Since these fixtures hang below the surface of a counter, they are prime candidates for under-the-rim leaks.

Molded lavatory tops

Molded lavatory tops have integral lavatory bowls and normally are made with a backsplash. This type of lavatory is the least likely to produce leaks that are not plumbing related. The only place these units are likely to leak is around the faucet holes. All lavatories have the potential to leak through these holes if their faucets are not properly sealed and mounted.

Troubleshooting Lavatories

Troubleshooting lavatories is not very different from working with any type of sink, but the pop-up assembly does allow for some additional problem possibilities.

The slip-nut connections and traps for lavatories should be checked out with the same procedures described for kitchen sinks. Remember to always fill the bowl to capacity and release all the water at once to check for drainage leaks. It can be helpful to use toilet tissue to rub around possible leak locations. The paper will expose dampness that you might not see otherwise.

Pop-up assemblies

Pop-up assemblies (Figs. 4.7 through 4.13) are the primary difference between lavatories and sinks. If a pop-up plug is not properly adjusted, the bowl will not hold water. This problem is usually obvious, and it is normally corrected by adjusting the lift rod. There are, however, times when a lavatory bowl will lose its water even when the pop-up plug is sealing properly. How can this be? Do you remember the situation I described in the opening of this chapter? I told you that a lavatory could lose all of its water and never show a trace of a leak. Well, you are about to find out how that can happen.

I have seen good plumbers struggle with pop-up adjustments over and over again when the pop-up plug was not the cause of the problem. Common sense tells you that if water is leaving the bowl, it must

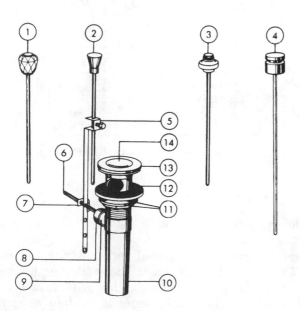

Figure 4.7 Parts detail for a lavatory pop-up assembly. (*Courtesy of Delta Faucet Company.*)

KEY	PART DESCRIPTION	ORDER NUMBER	RP5651	RP5651 BC	RP5651 CB	RP5651 CG	RP5651 G	RP5651 SG	RP6266	RP7526	RP7526 BC	RP7526 CG	RP7530	RP10109 BC OAK	RP10109 CG ROS	RP10109 OAK	RP10109 ROS	RP12498
1.	Knob & Rod Assy.-Crystal	RP7524								●								
		RP7524 BC									●							
		RP7529										●	●					
2.	Knob & Rod Assy.-Brass	RP6146	●															
		RP6146 BC		●	●													
		RP6146 G					●											
		RP6146 SG						●										
		RP6282				●			●									
3.	Knob & Rod Assy.-Wood	RP10108 BC OAK												●				
		RP10108 CG ROS													●			
		RP10108 OAK														●		
		RP10108 ROS															●	
4.	Knob & Rod Assy.-Brass	RP12501																●
5.	Strap & Screw	RP6136	●	●	●	●	●	●	●	●	●	●	●	●	●	●	●	●
6.	Horizontal Rod w/Clip	RP6134	●	●	●	●	●	●	●	●	●	●	●	●	●	●	●	●
7.	Spring Clip	RP6144	●	●	●	●	●	●	●	●	●	●	●	●	●	●	●	●
8.	Nut	RP6132	●	●	●	●	●	●	●	●	●	●	●	●	●	●	●	●
9.	Pivot Seat & Gasket	RP6130	●	●	●	●	●	●	●	●	●	●	●	●	●	●	●	●
10.	Tailpiece	RP6128	●	●	●	●	●	●	●	●	●	●	●	●	●	●	●	●
11.	Nut & Washer	RP6140	●	●	●	●	●	●	●	●	●	●	●	●	●	●	●	●
12.	Gasket	RP6142	●	●	●	●	●	●	●	●	●	●	●	●	●	●	●	●
13.	Flange	RP6126	●		●					●						●	●	●
		RP6126 BC		●							●			●				
		RP6126 CG				●						●			●			
		RP6126 G					●											
		RP6126 SG						●										
		RP6268							●				●					
14.	Stopper	RP5648	●		●					●						●	●	●
		RP5648 BC		●							●			●				
		RP5648 CG				●						●			●			
		RP5648 G					●											
		RP5648 SG						●										
		RP6313							●				●					

Figure 4.8 Parts description for the pop-up assembly in Fig. 4.7. (*Courtesy of Delta Faucet Company.*)

be getting past the pop-up plug. While this is usually the case, there is an exception.

If the rim of a lavatory drain is not sealed with putty, water can leak under the rim. We've talked about this in the section on kitchen sinks, but the leak is harder to find when dealing with lavatories. When the leak occurs on a kitchen sink, water is evident at some point beneath the sink. This is not the case with a lavatory.

Water that seeps under the rim of a lavatory drain winds up in the drain assembly. The water is carried down the tubing and into the

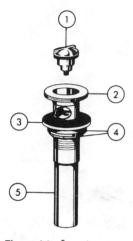

Figure 4.9 Lavatory pop-up with a rotary stopper. (*Courtesy of Delta Faucet Company.*)

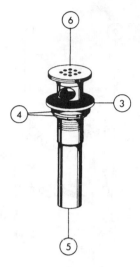

Figure 4.10 Lavatory pop-up with a grid strainer. (*Courtesy of Delta Faucet Company.*)

KEY	PART DESCRIPTION	ORDER NUMBER	MODELS B RP6339	C RP6346
1.	Rotary Stopper	RP6354	●	
2.	Flange	RP6126	●	
3.	Gasket	RP6142	●	●
4.	Nut & Washer	RP6140	●	●
5.	Tailpiece	RP6128	●	●
6.	Grid Flange	RP6344		●

Figure 4.11 Parts description for the pop-up assemblies in Figs. 4.9 and 4.10. (*Courtesy of Delta Faucet Company.*)

trap, without leaving any evidence of a leak. The only clue available is the knowledge that the bowl is losing its water. A plumber who is not aware of this type of possibility could work for days adjusting a pop-up assembly without success in solving the puzzle.

How can you tell if the problem is a pop-up adjustment or a bad drain seal? The easiest way to make a definite determination is to

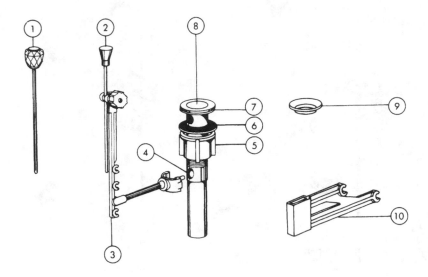

Figure 4.12 A Snap'n Pop pop-up assembly. (*Courtesy of Delta Faucet Company.*)

MODELS

KEY	PART DESCRIPTION	ORDER NUMBER	D RP6463	RP7525
1.	Knob & Rod Assy.	RP6146	●	
2.	Knob & Rod Assy.	RP7524		●
3.	Pivot Rod & Strap Assy.	RP6477	●	●
4.	"O" Ring	RP6481	●	●
5.	Nut	RP6475	●	●
6.	Gasket	RP6142	●	●
7.	Foam Gasket	RP6876	●	●
8.	Stopper	RP5648	●	●
9.	Spacer	RP10994		
10.	Extender	RP7585		

Figure 4.13 Parts detail for pop-up assembly in Fig. 4.12. (*Courtesy of Delta Faucet Company.*)

remove the pop-up plug and install a rubber stopper in the drain opening. If you fill the bowl with the rubber stopper in the drain and still lose the water, you can bet the problem is a bad drain seal. If the rubber stopper holds the water in the bowl, you have a pop-up assembly that is out of adjustment or a bad seal on the pop-up plug. This simple test can save you hours of time and frustration.

When you do the rubber-stopper test, use only a rubber stopper. If you use a rubber disk, such as is sometimes used to seal kitchen sink drains, the disk will cover the drain rim, and you won't get accurate data.

If you find the cause of your leak to be a bad drain seal, you must disassemble the drain assembly and install a new seal of putty.

When you know the problem lies with the pop-up plug, you can adjust the lift rod to compensate for the poor fit. If the leak persists, check the seal on the pop-up plug. If the gasket is bad, either replace it or the entire pop-up plug.

The threads of a pop-up assembly can allow water to leak by if they are not coated with a sealant. This type of leak will not normally show up when water is just being run out of the faucet. To expose this type of leak, the lavatory bowl should be filled and all the water released at once. You need the volume and pressure this type of test provides to find leaks in the threads.

Sometimes water will leak out around the retainer nut that holds the pivot rod of a pop-up assembly in place. This usually indicates a bad washer or the absence of a required washer. If you remove the retainer nut and pivot rod, there should be a rubber washer installed in front of the nylon ball. If the washer is missing, cut, or worn, a leak can occur.

Slow drains

Slow drains are common for lavatories. Human hair is routinely found to be the cause of these sluggish drains. Unlike most sinks, dropping a trap and inspecting it may not reveal the reason for a lavatory draining poorly.

Hair is commonly caught on the base of pop-up plugs and around pivot rods. A plumber who only inspects a trap and then opts to snake the drain of a lavatory may be wasting a lot of time and energy. The root cause of the problem may be hidden in the pop-up assembly.

Before you start snaking a slow lavatory drain, check the pivot rod and pop-up plug for hair and other obstructions. If the sink is not equipped with a pop-up plug, look closely in the drain assembly for unwanted objects, such as toothpaste container tops.

Open lavatory drains frequently collect tops from toothpaste containers and shampoo bottles. Coins and a wide variety of other unexpected items can be found in the drain assemblies of lavatories. To check for these items, remove the trap. Shine a light down the drain and do a visual inspection. The pivot rod may have to be removed in order to clear obstructions in the drain assembly.

Figure 4.14 A galvanized steel drain clogged with grease and other debris.

Another possible cause for a sluggish lavatory drain is a pop-up plug that is not being lifted high enough by the lift rod. If the lift rod does not protrude high enough above the lavatory faucet, it may not provide adequate leverage to give a full opening below the pop-up seal. When this problem occurs, a minor adjustment in the height of the lift rod will correct it.

Lavatories that are served by galvanized drain pipes require frequent attention. The rough inner surface of the galvanized pipe catches hair and other substances (Fig. 4.14), causing the drain to flow slowly. Additionally, the short-turn fittings used with galvanized pipe do not result in fast-flowing drains. We will go into more details on the drains and vents of plumbing systems in Chap. 18.

Common Sense

Common sense and a systematic approach will go a long way in any type of troubleshooting. This is particularly true when working with sinks and lavatories. There are very few complex reasons for a sink or lavatory to leak.

I have given you examples of the two most common hard-to-find leaks when dealing with sinks and lavatories—a lack of putty around the drain rim and a lack of putty around the fixture rim. Aside from these two types of problems, most sink and lavatory leaks are easy to find. It is, however, important that you remember to fill the bowls to capacity and release all of the water at once when testing for drainage leaks. Also remember to use paper towels or toilet tissue to help locate tiny leaks. If you remember these principles and troubleshoot sinks and lavatories with a good flashlight and a systematic approach, you shouldn't have much trouble finding the causes of leaks.

We are now ready to move onto the next chapter and investigate troubleshooting techniques for bathtubs.

5

Troubleshooting Bathtubs

Problems with bathtubs can cause a lot of damage. If you've done much service work, you know how fast a bad bathtub can ruin a downstairs ceiling or floor. If you're new to the trade, you will soon find out how destructive the leaks associated with bathtubs can be.

Because of the nature of bathtubs, they can offer plenty of challenge to a troubleshooter. There are many ways that leaks having to do with bathtubs can avoid detection or cast suspicion on other parts of a plumbing system.

Bathtubs can produce leaks from their wastes and overflows, tub spouts, shower heads, valves, traps, and even their surrounding walls and floors. With so many possibilities to consider, troubleshooting bathtubs requires strict attention to detail and a plan of action.

Let's start this chapter with a look at a real-life situation I have encountered numerous times. The example you are about to read is not uncommon; it is, in fact, fairly typical. The story is based on actual service calls I've been on. The actions of the plumbers in the story mirror the actions of some of my past plumbers.

The story begins with a panicked homeowner calling for emergency service. She is calling from her kitchen phone as she watches a steady stream of water dripping from the dining-room chandelier. A plumber is dispatched to the location and arrives quickly. When the plumber enters the house, there is plenty of evidence that there has been a leak, but the water has stopped running out of the light fixture. The dining-room table is wet, the carpet is wet, and the homeowner is baffled by the leak disappearing before the plumber arrived.

To give you some background information, there is a full bathroom and a half-bath on the second floor of the house. Neither of the bathrooms is in the vicinity of the dining room, where the leak was noticed. Can you solve the puzzle yet?

After calming the homeowner, the plumber removes the light fixture's trim ring and shines a light into the ceiling. There is water standing on the back of the drywall ceiling. The puddle is surrounding the electrical box, giving no clear evidence of what direction it came from. With the limited viewing space available, the plumber cannot tell if there are any pipes in the joist bay. What would you do first in this situation? Before I tell you what the proper troubleshooting procedure is, let me explain how many plumbers would proceed.

A high percentage of plumbers would insist on cutting a hole in the ceiling to investigate the leak. After ruining the ceiling, these plumbers would probably not know much more than they did before opening the ceiling. It is unlikely that there would be any pipes in the joist bay, but they might find a trail of water to give some direction of where the leak originated.

While it might ultimately be necessary to cut the ceiling open, it would be a mistake to do so at such an early stage of the troubleshooting process. The plumber would be justified in cutting the ceiling, but there probably is a better way to reveal the cause of the leak.

A plumber who cut the ceiling open and found a trail of water would know what direction to go in but not how to proceed. Some plumbers would go to another point on the ceiling and make another inspection hole. This gives the homeowner two holes in her ceiling, and the plumber is probably no closer to resolving the problem.

What data do we have to assess so far? We know there are two bathrooms upstairs and that there are no pipes in the joist bay where the water ran through the light fixture. We know a significant volume of water leaked out of the ceiling, but then it stopped as quickly as it had started. What does this tell us?

We can tell now that the leak is not likely to be in the water distribution pipes. If it were, the leak would continue. The constant water pressure in the distribution pipes would not allow the water to cease.

Since the leak is not in the water pipes, it must be in the drainage pipes. The question is, what drainage pipes? There are no drains visible from the inspection holes in the ceiling. Many plumbers would be stuck at this point, not knowing what to do next. Some would probably hack out chunks of the ceiling until they found the leak, but they may still never find it. Leaks of this nature, especially if they haven't been occurring for an extended time, can be very difficult to pinpoint.

The water could have come from a toilet, a bathtub, a broken drain pipe, a faulty roof flashing, a water spill on an upstairs floor, or some other equally difficult circumstance. Since this is the first time such a leak has occurred in the house, there is no pattern to follow. How would you proceed?

Okay, enough of the suspense; let me tell you the right way to troubleshoot this problem. Since the water stopped on its own, it is safe to assume the leak is not from a pressurized water pipe. Before cutting the ceiling open, it would be wise to cut the power off to the light fixture and try to re-create the leak. If the leak showed up once, it will probably show up again. But, there are times when making the leak recur will seem impossible; I'll explain in a few moments.

When you assess the leak, you can see the volume of water is greater that what should have been produced by a lavatory. The amount of water present leads you to believe it must have come from a toilet, shower, or bathtub. Unless it has been raining, you can rule out the possibility of a leaking roof flashing.

The first step would be to ask the homeowner to stand watch at the light fixture. You should explain what you are going to do, and ask the homeowner to notify you if she sees water anywhere.

Ask the homeowner if anyone had used any of the upstairs plumbing within 30 minutes of the time the leak was discovered. If the plumbing had been used, and you can bet it had been, ask what fixtures were used. Did someone take a shower? Did someone take a bath? Was a toilet flushed? After getting answers to these questions, determine which bathroom was used, and start your troubleshooting in that bathroom.

We will assume that both bathrooms were used as the family was getting ready for work and school. All of the fixtures were used, and the leak was not noticed until the homeowner came downstairs.

Go upstairs and flush one of the toilets. Flush the same toilet at least twice. Wait a few minutes before moving onto the next fixture. Since the water is traveling across the ceiling before coming out, it will take a little time for it to reach the light fixture.

Once you are convinced the first toilet flushed is not the cause of the leak, flush the second toilet twice. After waiting for the water to show up at the light fixture, test the bathtub. Fill the tub full of water and then open the drain. Let all the water drain and wait to see if it shows up at the light fixture. If it doesn't, try running the shower for 5 minutes.

The riser to the shower head could have a leak in it. This leak wouldn't show up unless the shower was turned on; the riser is not under pressure unless the shower is running. If you still haven't been able to make the water reappear at the light, try the lavatories. Fill each lavatory and let them drain, one at a time.

In many cases you would have found the leak by now, but this is not an average job. You've tested all of the plumbing and nothing has caused water to come out of the light fixture. What should you do now? Well, there are still a number of things to check out.

If none of the plumbing produced the leak, it had to come from a less-obvious place. We will assume that the full bath is equipped with a bathtub that has a shower head over it. The bathroom floor is tiled and so are the walls that surround the bathing unit.

Where might the water have come from? It could have come through the floor or the tub walls. If the grouting in the tile work has gone bad, the water may have passed right through and worked its way over to the light fixture. If the tub spout or tub faucets are not properly sealed, the water may have splashed off the bather and run down into the wall cavity and onto the ceiling.

Since these types of leaks won't show up simply by running the shower, you must simulate the conditions when someone is taking a shower. No, you don't have to stand in the tub and get splashed. If the tub is equipped with a shower curtain, you're in luck.

A shower curtain is a good sign that water escaped from the bathing unit, onto the floor. The curtain can also be used to simulate the splashing that occurs when someone showers.

Start this phase of your troubleshooting by pouring water on the bathroom floor, particularly near the edge of the bathtub. Don't skimp on the water. If the water disappears, expect to hear the homeowner screaming soon. Should the water remain on the floor, move your investigation to the tub walls.

Turn on the shower and hold the shower curtain toward the front of the tub, in about the same place a bather would stand. Let the water bounce off of the shower curtain and onto the front and side tub walls. Keep this up for at least 5 minutes. When the homeowner lets you know the leak is back, you have narrowed the list of potential causes. It is either bad grouting or poor seals around the tub spout or faucet handles.

Have the homeowner come upstairs and watch through the bathtub's access door to see if the water is running through at the plumbing outlets or if it is coming through the grouting. If it is not pouring past the plumbing holes, you can assume it is running through the tile walls.

As you can see, this type of troubleshooting will be time consuming. If you have a good eye for construction, you may be able to speed up the process. For example, I can usually look at a ceiling and tell which way the floor joists are running. Since water normally doesn't cross through joist bays, I can often pinpoint precise locations for detecting the leak.

Tubs That Won't Hold Water

Tubs that won't hold water generally require an adjustment to the tub waste. It is possible for water to leak around the trim ring of the tub drain, but the problem is usually with the adjustment of the tub waste. If there is no water leaking from under the tub, you can rest assured the cause is in the tub waste adjustment.

There are several different types of tub wastes in use, and each of them can present its own type of trouble. With some types of wastes it is necessary to make adjustments in the trip mechanism. Others require the replacement of a seal or an entire stopper.

Toe-touch wastes

Toe-touch wastes don't contain any internal adjustment mechanisms. This type of waste depends on the stopper to hold water in a bathtub. If a toe-touch waste is allowing water to leak out of the tub and down the drain, there are three options available to you.

Begin your troubleshooting by turning the stopper clockwise. If the stopper is not screwed into the drain far enough, water can leak past its seal. If the stopper is screwed into the drain properly, remove it and inspect the seal. To be sure the problem is with the toe-touch stopper, install a rubber stopper in the drain and test the tub. If water doesn't leak past the rubber stopper, you have identified the cause of your problem; it is the toe-touch stopper.

Once you know the stopper is causing the trouble, you can replace the entire stopper with a new one. It may be possible to replace the seal only, but with the low cost of replacement stoppers, I would replace the whole unit to be sure of correcting the problem without a callback.

Lift-and-turn wastes

Lift-and-turn wastes (Fig. 5.1) can be treated like toe-touch wastes when troubleshooting for seeping water. Even though the two wastes don't operate in the same manner, they do seal the tub drain in the same basic way.

Mechanical strainer wastes

Mechanical strainer wastes (Fig. 5.2) are more difficult to deal with than toe-touch or lift-and-turn wastes. This type of waste depends on a plunger blocking the tee of a tub waste to maintain a static water level in the tub. Fine-tuning the adjustment of this type of tub waste can require several attempts.

To rule out a leak around the trim ring of the tub drain, you can remove the strainer and insert a rubber stopper. This test will tell you positively if the problem is with the plunger of the tub waste.

Once you know the plunger needs to be adjusted, you must remove

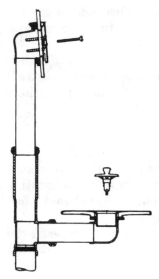

Figure 5.1 Lift-and-turn tub waste. (*Courtesy of Delta Faucet Company.*)

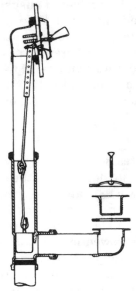

Figure 5.2 Trip-lever tub waste. (*Courtesy of Delta Faucet Company.*)

the screws from the overflow plate. This allows you to remove the entire trip mechanism. The plunger location can be raised or lowered by turning the threaded portion of the mechanism. Since water is leaking past the plunger, the length of the mechanism should be extended.

Reinstall the adjusted mechanism and test it. It is not uncommon to have to tinker with the adjustment several times before the perfect position is found.

Rocker-type wastes

Rocker-type wastes require similar troubleshooting to that of strainer wastes. A rocker waste, however, has a spring, rather than a plunger, that hangs in the overflow tube. The spring comes into contact with the rocker assembly and allows the tub stopper to be opened and closed.

The troubleshooting procedure for this type of waste should begin with the rocker assembly. Operate the trip lever and observe the stopper. Does it seat into the drain evenly? If not, the position of the rocker assembly should be adjusted. Does the stopper drop deeply into the tub drain? If not, the height of the stopper should be lowered.

It is not unusual for the rocker assembly to be the cause of seeping water with this type of waste; however, the problem could be with the trip mechanism. If the stopper doesn't pop up and seat with satisfactory action, the length of the trip mechanism may require adjustment. This is accomplished in the same way that was described for a strainer waste.

Tub-shoe leaks

Tub-shoe leaks can be the cause of water draining from a bathtub inadvertently. When water seeps past a tub-waste mechanism, it goes down the drain. But when the leak is at a tub shoe, the water goes into unwanted areas, such as a downstairs ceiling.

A lack of putty on the underside of the tub drain is the most common cause for tub-shoe leaks. It is possible, however, that the leak is due to a faulty or missing tub-shoe gasket. I have worked on tubs where the tub-shoe gasket was never installed.

If you are faced with a situation where there is no access to the tub shoe, you must do your work from inside the bathtub. This is really not difficult, and it saves time and trouble of cutting out a ceiling. All that is required is the removal of the tub drain.

Once the tub drain is unscrewed, the gasket can be retrieved and inspected. Replacing the gasket requires a little manual dexterity, but it can be done from inside the tub without much trouble.

I sometimes put a little pipe dope or wax on the back of the gasket to make sure it doesn't slide off the tub shoe before the drain is replaced. Losing your only gasket between the floor and ceiling can be embarrassing, and the tacky dope or wax minimizes the risk of loss.

By installing a new gasket, putting sealant compound on the drain's threads, and installing putty around the inner rim of the drain, you can be fairly certain of stopping a leaking tub shoe.

Bathtubs That Drain Slowly

Bathtubs that drain slowly are a common problem. The cause of the problem is usually either a tub waste that needs to be adjusted or a partial blockage in the drainage system. A vent that is obstructed or nonexistent can also contribute to the slow draining of a bathtub.

Hair is a frequent enemy of bathtub drains. Even tubs equipped with strainer-type drains can be slowed by accumulating hair. Mechanical tub wastes are particularly susceptible to hair attacks. Let's look at the different types of wastes and see what to do when a tub is not draining fast enough.

Toe-touch wastes

Toe-touch wastes are easy to troubleshoot. You can unscrew the stopper and inspect the crossbars in the drain for obstructions. If there aren't any, you should try installing the stopper so that it sits a little higher than it did before. The simple act of loosening the stopper may be all that is required to solve the problem of sluggish draining.

You can either identify or eliminate the stopper as the cause of your problem with ease. Put a rubber stopper in the tub drain, and fill the tub partially with water. Remove the stopper, and observe the draining action. If the water leaves the tub quickly, the stopper is your problem. When the water is slow leaving the tub through a wide-open drain, you must look further.

It is possible, but unlikely, that an obstruction has lodged in the tee of the tub waste. A small spring-type snake will eliminate this possibility quickly.

Lift-and-turn wastes

Like with a toe-touch waste, you can remove the stopper of a lift-and-turn waste to inspect the drain's crossbars for obstacles. While the stopper is out of the drain, cover the drain opening and test the tub as described above. There isn't much in either a toe-touch or lift-and-turn waste to cause trouble.

Mechanical strainer wastes

The plunger used with a mechanical strainer tub waste is frequently a cause for slow drainage. The plunger may be engulfed with hair, or it may simply be out of adjustment.

Remove the overflow plate and trip assembly. If the plunger is free of debris, there are only two likely causes for slow drainage. Either the plunger needs to be raised by shortening the lift assembly, or there is a blockage somewhere else in the drainage system.

With the plunger removed from the overflow tube, cover the drain and test the tub as described earlier. If the water drains freely, the plunger needs to be adjusted. Should the water remain sluggish without the plunger installed, look for a drainage obstruction.

By testing the tub drain with the plunger removed, you can save yourself a lot of tedious trouble. I've had plumbers spend close to an hour adjusting the height of plungers only to find out that the problem was not with the plungers at all. It pays to work smarter, not harder.

Rocker-type wastes

Rocker-type wastes can catch hair on the rocker assembly and the spring that operates it. Grasp the stopper and remove the rocker assembly. If it is not fouled with hair, remove the lift assembly from the overflow tube. With both of these parts removed, cover the drain and test the tub as described earlier.

If water runs out of the tub quickly when the waste components are removed, and the rocker and spring are clean, you can plan on having to adjust the tub waste.

Trap and Drain Obstructions

When a tub waste is not the cause of a slow-draining bathtub, look for trap and drain obstructions. In most cases this can be done by running a small spring-type snake down the overflow tube of the tub waste. A 25-foot snake should reach well into the drainage system and eliminate any obstruction that is affecting only the bathtub.

Most bathtubs are fitted with P-traps, which make snaking them easy. If you encounter a drum trap, which is illegal for most applications, you must gain access to the trap and a cleanout opening. Otherwise, your snake is not going to get past the trap, and the cable is likely to kink.

Because of their design, drum traps are impossible to snake through from an overflow opening of a tub waste. The snake can get into the trap, but it cannot get out. At first, the snake will seem to be going someplace, but all it is doing is curling up inside the trap. If you are using an electric drain cleaner, the cable will probably kink, possibly to the point where it is damaged permanently.

There are a lot of drum traps in Maine. In fact, I was working with one just a few days ago. I was called out to a tub that wouldn't drain at all. Even after I removed the interior parts of the tub waste, the fixture wouldn't drain.

Maine's plumbing systems are not vented very well, so plungers often work to clear stoppages, but not on this job. I covered the overflow hole with one plunger and plunged the drain with another. The plungers had no affect on the stopped-up drain. I was relatively certain the tub would be equipped with a drum trap, but I snaked the drain as far as I could. The cable started to kink, and I was sure there must be a drum trap on the tub drain.

There was a closet behind the head of the tub, but there was no access panel. With a mechanical tub waste and a drum trap, there should have been an access door in the wall, but there wasn't. To avoid cutting into the closet wall or the ceiling down below, I tried to clear the stoppage with water pressure from a balloon bag. Unfortunately, the results were not good. The tub started to drain ever so slightly, but it was still clogged up.

The property was a multifamily building being used as rental property. I called the property owner and got permission to cut a hole in the closet wall. When I did, I was not surprised to find a drum trap.

The trap was an old one, with a brass cleanout in it. The piping was all galvanized steel, and there were three fittings within a foot of the trap. There was really little wonder that the drain wasn't working well.

The trap was hanging between two floor joists. With the limited space between the joists, I had no room to get leverage on the cleanout with a pipe wrench. I tried to remove the cover with a basin wrench, but the cover was too tight to budge.

This type of problem is not unusual when you find an old drum trap on a bathtub. How would you get the cleanout cover off? I tried using a socket, socket extension, and rachet to remove the cover, but it still wouldn't turn. I hit the cleanout with a hammer to loosen it, but still no dice.

The next logical step was to take a cold chisel and beat the cover out of the trap. I didn't want to do this, but the limited space I had for access wasn't giving me much of an alternative. Just before I started banging away at the cleanout, I got an idea.

I dug around in my truck and found a long handle that would fit on the socket extension. With the use of the long handle, a lot of muscle, and a little luck, I was able to get the cover off the trap.

When I put the snake in the drain pipe, it hit resistance almost immediately. There were two galvanized eighth-bends on the discharge side of the trap. Only a short nipple separated the two fittings, and the clog was in the offset. A few good turns of the snake and the drain was clear.

The job turned out all right, but it could have gotten nasty if I hadn't known what to do about the drum trap. I could have fought with my snake for quite awhile, thinking it was hitting the stoppage when, in fact, it was hitting the drum trap.

If I hadn't had the socket set and long handle, I might not have been able to remove the cleanout cover. Experience also played another vital role in this particular job. The drum trap had been installed upside down, so the cleanout was on top. If I had opted to cut the ceiling from below, instead of the closet wall, I still wouldn't have had access to the cleanout.

Why did I choose to open the wall in the closet? There were two reasons. I suspected the trap might be installed upside down because I have found many like it in the past. Secondly, I knew the damage to a closet wall would not be as disruptive for the property owner as a gaping hole in the downstairs tenant's apartment. These are the types of thoughts you should consider when troubleshooting and deciding on a course of action.

Vent Blockages

Vent blockages can cause a fixture to drain slowly. If you have checked the tub waste and the drainage system without finding a cause for slow drainage, turn your attention to the vent system. We are going to go into great detail on vents and venting in Chap. 18, but it is appropriate that we take a look at how they affect bathtubs in this chapter.

Some bathtubs have individual vents, and others are either wet-vented or act as wet vents. And then, there are the tubs that are not vented at all.

If a tub has an individual vent and is draining slowly because of the vent, the problem is usually that the vent has become obstructed by leaves or other debris that has entered it. Snaking the vent pipe from the roof or attic of the property solves this type of problem.

Bathtubs that are wet-vented are usually vented through a lavatory drain. Under these conditions the vent to check out is the one near the lavatory. This type of setup is not unusual.

Tubs that serve as wet vents for other fixtures, such as toilets, can basically be treated as if they were individually vented. These tubs have individual vents, so check the vent close to the tub.

A sign that a tub vent is not working properly is a gurgling noise in the drainage system. The sound is similar to the one made when a drain is stopped up. If you have investigated all aspects of the drains and can't find a reason for them to be running slowly, take a close look at the vents.

It will be helpful to ask the property owner if the fixture has always drained slowly. If the bathtub has never drained properly, it may not be vented at all.

Overflow Leaks

Overflow leaks can be difficult for inexperienced plumbers to identify. This type of leak will not show up through normal test procedures. Overflow leaks occur when a tub is filled to capacity and someone gets into the water. The displacement of water caused by the human body can create an overflow leak.

The overflow fitting of a tub waste should be equipped with a tapered sponge gasket. This gasket prevents water that is going over the overflow plate from leaking past the drainage system. Sometimes the gasket is installed upside down, and there are occasions when the gasket is never installed. It is also possible that the gasket has lost its holding ability and is allowing water to sneak past it.

If there is an access panel at the head of the tub, an overflow leak is easy to identify. If the leak has been going on for some time, there should be evidence of water in the access opening. The leak may, however, be running through the hole that was cut in the floor to allow the installation of the tub waste.

How do you find an overflow leak? Well, you could fill the tub and get in it, but there is an easier way. Fill the tub to capacity and place a heavy object, like a 5-gallon bucket filled with water, in the bathtub. The mass will displace the water and force it to run out the overflow as it does when someone is bathing. By watching the overflow system from the access panel, you will be able to spot the leak.

If you are working alone and can't make the tub overflow without being there to hold pressure on the displacement object, you must get a little more creative. Wrap toilet tissue or a paper towel around the overflow tube. Spread out some more on the floor around the overflow. Then go back to the tub, and force the water to overflow. Any leak will show up on the paper towels and toilet tissue.

Escutcheon and Tub-Spout Leaks

Escutcheon and tub-spout leaks can be extremely hard to find unless you know to look for them. When escutcheons and tub spouts are not installed with the proper gaskets or putty, they can allow water to run right past them. The water can get into the wall and drip down to the floor. The leak can spread out and be difficult to pinpoint.

A garden hose is ideal for testing this type of situation. By spraying water on the escutcheons and tub spout, you can determine if they are leaking into the access area. If you don't have a garden hose to use, fill a pitcher or bucket with water and splash it on the escutcheons and spout. You can use paper towels in the access area to help locate any stray water.

Escutcheon and tub-spout leaks only occur when water is splashed against them. The most common cause is when someone is taking a shower and the water bounces off his or her body and onto the walls.

Valve Leaks

Valve leaks can be detected easily when there is an access panel to work with. Turn on the hot and cold water and inspect the valve. Use toilet tissue to wipe joints and discover small leaks.

Don't overlook the possibility that the valve body may be pitted or cracked. This condition occurs more often than you might think. Remember, when troubleshooting, to look for every possible problem. Inexperienced plumbers and plumbers who are in a hurry often overlook the obvious.

I've installed brand-new valves that had tiny holes in the bodies.

These leaks are not always easy to find, and over the course of time, they can cause a lot of damage.

If you are troubleshooting an existing valve or replacing an old one with a new one, don't fail to inspect it closely for hairline cracks and invisible holes. Sometimes these leaks are so small that they don't even really drip. Water seeps out of the openings and spreads across the valve until it has accumulated enough to drip off the valve. If the water happens to slide down the valve and onto a supply pipe, it can easily go undetected.

Using toilet tissue to find these little leaks is the most effective way to ensure success. If you do not look for these minuscule leaks, trouble is sure to find you in the future. Once the valve is concealed in a wall, the leaks will eat away at the building materials a little at a time, for a long time to come. Then one day, the property owner will notice a damp carpet or a stained ceiling. The evidence may even show up in the lower sections of the tub wall. However the leak is found, the damages that have resulted, and they can be extensive, may be found to be your fault. Don't let this happen to you; check your work closely for leaks.

If you don't have open access to the tub valve, you can do a limited visual inspection through the holes in the wall where the valve stems protrude. Remove the handles and escutcheons, and shine a light into the cavity. Test the valve with the water both on and off. As a last resort, make an access panel to inspect the back side of the tub.

The Process of Elimination

The process of elimination is used in all forms of troubleshooting, and it should be used with bathtubs. There are very few elements of a typical bathtub that can cause problems, but putting your finger on the right cause can be difficult.

Many types of tub leaks can appear to be something they are not. For instance, water leaking past a tub spout and dripping down to the floor can look like a leak in the waste and overflow. Water that runs past the overflow gasket can saturate the floor without leaving a trace of where it came from.

Plumbers don't normally get called in on hidden tub leaks until some damage has been done. By the time a plumber arrives, water has often traveled away from the leak, making the leak harder to find.

Using the proper troubleshooting procedures and a process of elimination will reveal the leaks. It may take some time and patience, but the leaks can be found.

As you have seen, there are several types of leaks associated with bathtubs that will not show up in typical testing. I have shared my many years of experience with you to illustrate simple ways to eliminate possible causes of trouble. If you move systematically through the troubleshooting process, and remember to look for the hidden causes, you shouldn't have any difficulty troubleshooting bathtubs.

6

Troubleshooting Showers

Showers come in all shapes, sizes, and types, but they all share the same basic characteristics. Because of their design, showers don't harbor a lot of potential for baffling problems. This is not to say, however, that showers can't cause a lot of trouble. While showers don't offer many possibilities for problems, they can be difficult to troubleshoot and to work on.

Unlike bathtubs, where an access panel allows some visibility to the drainage area, the traps and drain pipes for showers are usually inaccessible. Pan-type showers can be a plumber's worst nightmare, and shower curtains can create some perplexing problems.

What Can Go Wrong with a Shower?

There aren't a lot of plumbing problems associated with showers, but the damage that can be done by a shower can be extensive. If a shower pan is leaking, it can be confused with a leaky drain. A shower arm with deteriorated threads can avoid detection from all but the best of plumbers. Showers with tile walls can present plenty of challenge for finding the cause of a leak, and shower doors and curtains can confuse the issue of troubleshooting.

Many water problems with showers are related to the users of the showers. While these problems are not pure plumbing problems, plumbers are responsible for finding the causes. Let's move on now to specific situations and see how you can improve your troubleshooting effectiveness when dealing with showers.

One-Piece Showers

One-piece showers offer the least possibly for problems of all the various types of showers. Because of their one-piece construction, leaks in the walls are nearly unheard of. Unless the unit is cracked, you can rule out wall leaks.

Valve Escutcheons

The valve escutcheons on showers are one source of mystery leaks. Just as was described in the last chapter, these escutcheons can let water penetrate the shower wall.

The escutcheons used for single-handle valves are frequently equipped with sponge-type gaskets. These gaskets are factory installed on the escutcheons and do a good job of preventing unwanted leaks.

Two-handle shower valves generally require a gasket or a putty seal to be placed behind the escutcheons. If these seals are missing or have deteriorated, water can find its way into the wall. A visual inspection can reveal the possibility of this type of leak, but a water test is the only way to be certain the escutcheons are not responsible for leaks.

Unfortunately, many showers don't have access panels that allow easy inspection of the shower valve. Bathtubs that use waste and overflows with mechanical joints are required to have access panels, but showers are not.

The lack of an access panel greatly reduces a plumber's ability to make sound judgment calls on problems related to showers. It may be necessary to cut into a wall to gain access in the case of stubborn leaks, but there are ways to make reasonable assumptions without butchering a wall. To expound on this, let's look at various circumstances on a step-by-step basis to see how cutting into walls and ceilings can be avoided.

Escutcheons

Escutcheons are notorious for creating hard-to-find leaks. When you combine the lack of an access panel with a leaky escutcheon, you have a tough situation to troubleshoot. Many plumbers would assume a wall would have to be cut to determine positively that an escutcheon is allowing water to pass through the wall. This is a logical assumption, but there is an alternative that will often work.

How can you tell if water is leaking into a wall through an escutcheon if you can't see the leak? Remove the valve handles and escutcheons. Pack the hole around each valve stem with toilet tissue, and replace the escutcheons. Spray or splash the escutcheons with water, and then remove them again and check the toilet tissue. If the paper is wet, you know the escutcheons are leaking. Dry paper, of course, indicates no leak at the escutcheons.

The toilet-paper test is not an ideal way to troubleshoot escutcheon leaks, but if you can't do anything else, it works pretty well.

Packing putty behind the escutcheons should block water, but you may also want to run a bead of caulking around the outside edges of all the escutcheons.

Shower Arms

Shower arms can rot walls, floors, and ceilings. Houses that derive their water from wells are especially susceptible to this problem.

Too many plumbers fail to inspect the inside-the-wall connection of a shower arm. I guess it's the old out of sight, out of mind thing. If you are faced with a shower leak you cannot find, check the connection between the shower arm and the shower ell.

How do you check a shower-to-ell connection without an access panel. It's really quite simple; pull the arm escutcheon forward and look through the hole. Remember to make sure water is running through the shower head when you do the inspection. It's hard to believe, but I've seen plumbers inspect shower risers and their connections without cutting the diverter on. If water is not coming out of the shower outlet, you are not going to find a leak in the shower riser.

The threads of shower arms sometimes rust out, and that is mostly what you are looking for. It's also possible that the arm was never doped up before being installed or that it just isn't tight.

Shower Ells

Shower ells can sometimes spring leaks in their castings or at their connection with the shower riser. You can check for these hidden leaks in the same way that was described for checking shower arms.

Shower Risers

Shower risers have been known to spring leaks in the pipe material, the connections, and the fitting and valve castings. For example, water with a high acidic content will often eventually eat pinholes in copper tubing. I've seen this happen a number of times; however, I've never found it to be a problem with shower risers.

If there is a problem in a shower riser, it is usually at the connection point with the shower ell or the shower valve. You can inspect the shower ell connection by following the instructions above. It's usually possible to inspect the riser-to-valve connection through one of the holes surrounding a valve stem. Again, make sure the shower head is turned on when you look for the leak.

If you work in an area subject to freezing temperatures, you might come face to face with a shower riser that has frozen and split. This is very common when dealing with seasonal cottages. Let me give you two examples of this type of problem.

The first example involves a leak that is relatively easy to find. This is the case when a shower riser has frozen and split with a large gap in the pipe. About all you have to do to find this type of problem is turn on the water to the shower riser. You will hear water rushing in the wall.

When a shower riser splits, the wall must be opened to make the necessary repair. The good thing about this type of leak is that it is easy to find. A large volume of water is released and the leak exposes itself before much damage is done. The bad thing is that the wall must be cut open.

The second type of freeze-up leak is not so easy to find, and it can do a lot of damage before it is discovered. A lot can happen to copper tubing that freezes. The tubing can blow out of fittings, swell and spilt with large gaps, or begin to split and then thaw before a large hole is made. Large splits and blowouts are easy to notice, but small holes in shower risers can go unnoticed for a long time.

I said earlier that these types of problems were common when working with seasonal cottages. People who drain down the plumbing systems in these types of properties and winterize them sometimes neglect shower risers. Even professional plumbers sometimes fail to protect shower risers when winterizing properties.

If the person draining the plumbing system fails to open the shower valves or, in the case of a tub-shower combination, the diverter valve, water remains trapped in the shower riser. This is an easy mistake for a homeowner to make, but professional plumbers shouldn't fail to drain shower risers.

When a shower riser swells under freezing conditions and begins to split, the crack can be very small. If thawing temperatures occur before a pipe is split completely, the gap does not mature, but it will still leak.

Small splits in shower risers can leak for a long while before they are found. Since the risers are not under constant pressure, the leaks only spill water when the shower is being used. If the hole in the pipe is small, the water will run down the piping and gather on the floor. This process will continue every time the shower is used. As time goes by, the water slowly rots flooring, walls, joists, and other building materials.

These tiny leaks don't make much noise, and they are downright difficult to find without an access panel. There is a way to test for small leaks in shower risers without cutting into the wall; do you know how to do it?

If you suspect a leak in a shower riser but don't want to cut open a wall until you're sure, you can put a pressure test on the riser. Turn the shower valves off, and remove the shower head. Screw a test rig on the shower arm, and fill the riser with air pressure. Leave the test on the pipe for 10 minutes, and then check the air gauge. If the riser has a leak in it, the pressure gauge will drop, and you can feel safe in opening the wall. This little trick of the trade can save you from the embarrassment of destroying a wall unnecessarily.

Valve Leaks

Valve leaks are not uncommon, and you can inspect for most valve leaks through the holes where the valve stems come out. It is possible that the back of the valve is leaking and that you won't be able to see it.

These suggestions will help you to avoid cutting walls unnecessarily, but there will be times when opening a wall is the only option left.

If a leak is found through the frontal inspection process, open the wall to correct the leak. The only exception to this is the shower arm; it can be corrected without destroying the wall.

Sectional Showers

Sectional showers have more risk of suffering from wall leaks than one-piece showers do. The reason is simple; they are not an integral unit. Any time there is a seam in something that is meant to be watertight, there is a risk of a leak.

Some sectional showers are better than others. There is no question that many high-quality sectional units are available, but there are also a number of low-grade showers on the market.

Even the best sectional shower can leak if it is not installed properly, and many plumbers never take the time to read and heed the manufacturer's installation instructions. Too many plumbers assume their generic knowledge is all that is needed to install a sectional shower. It is beneficial to always read over the installation instructions for a given shower prior to installation. If you, as the plumber, fail to make the installation in accordance with recommended procedures, you may be held liable for damages caused by leaks at a later date.

There are, to be sure, many types of sectional showers available. Some are two-piece models, some are three-piece units, some are screwed to wall studs, and others are glued to drywall. Each type of unit offers its own version of leak possibilities. Let's take a close look at each type and discuss what should be looked for when troubleshooting a sectional shower.

Adhesive applications

Adhesive applications are common among inexpensive shower surrounds. These three- to five-piece units range in price from about $40 to over $270. I have never installed one of the cheap versions, but I have seen a lot of them leak. On the other hand, my plumbers and I have installed hundreds of the upper-end models and never experienced any problems. What does this mean? Does it mean the cheap models are no good and the expensive ones are? Not necessarily.

The problems I have found in inexpensive glue-on surrounds have always been related to either the adhesive or the finish caulking. It may be that the lower-grade material shifts and stresses more than the more expensive surrounds do, or it may just be a matter of how the walls are installed.

With the type of adhesive surrounds I've used for years, the application of the adhesive and the caulking is of paramount importance, as it is with any shower walls of this type. I have personally trained my plumbers in the proper installation methods, and for the most part, they have followed my instructions. I did have a plumber once who installed the section with the integral soap dish upside down, not once, but twice in the same home. Needless to say, that plumber is no longer with my company.

If the proper adhesive is used in accordance with the manufacturer's recommendations, there is no reason why it should pull loose. Yet, this seems to be a common problem with the less-expensive surrounds.

The overlap between sections also plays a strong role in preventing leaks. The brand I use has a substantial overlap. This forces any water that penetrates the caulking to run down the inner lap to the shower base and out into the pan. Many cheap models offer only modest overlaps.

The flexibility and installation of caulking along the seams is of prime importance. If the installer doesn't get a good seal on the joints, a leak is likely.

If you have a sectional shower that is leaking through the floor, check the seams. If you find voids in the caulking, remove the old caulking and recaulk the entire seam. It would be a good idea to recaulk all the seams.

When you inspect a glue-on sectional shower, test the shower walls to see that they are attached firmly to the walls of the building. If the walls have pulled loose, water can slip behind them and cause problems. The result can be a rotted wall, a rotted floor, or a stained ceiling.

I had a shower once that was leaking to a ceiling below it, and I couldn't find the cause. I had checked all the seams, all the interior plumbing fittings, and ultimately the drain and trap. Even after cutting out a section of the ceiling for access to the bottom of the pan, I couldn't find a leak. I was not only frustrated, I was downright angry that the shower was beating me.

My work on the shower had started in the afternoon and went into the early evening. I wasn't willing to give up, but I was stumped. It was then that the husband came home and everything clicked. As soon as I saw him, I knew what I had been missing.

The man was very tall and was balding on the front of his head. I had noticed the shower head was set higher than normal, but I wrote that off to some plumber not following standard rough-in dimensions. However, when I saw the size of the man and put that together with the height of the shower head, I knew where the leak was coming from.

I went back upstairs and checked the top edge of the sectional shower surround. Sure enough, there was no caulking around the top edge. I surmised that when the man showered, water bounced off of him and onto the top edge of the shower.

After caulking around the top edge and asking the homeowners not to use the shower overnight, I waited to see what would happen the next day.

The hole in the ceiling was left open until we could determine if the problem was corrected. After a week of routine showering there had been no leaks. The homeowners waited another week to have the hole patched, and the shower never leaked. I had found and fixed the problem. That was the first and only time I've ever encountered such a leak, but you will do well to remember the circumstances in case you ever run up against such a problem.

Vertical screw-on surrounds

Vertical, screw-on surrounds are sometimes used as waterproof shower walls. These surrounds are usually made of good, heavy material, and since they snap into place and are screwed to wall studs, they tend to be solid.

Most models use vinyl strips to seal seams and hide the heads of screws. These strips can be worth investigating if you are fighting a frustrating leak.

The track that the sectional pieces sit in can also allow water to leak past the walls. If a panel is not put into place properly, there can be a void between the panel and the track. Caulking around this area normally prevents leaks, but the caulking can go bad, especially if it is not flexible.

Shower bases sometimes have to handle heavy weight loads, and the stress can pull caulking away from seams. If the property where the shower is installed is fairly new, the settling process of new construction can stretch the caulking. Always investigate the caulking on sectional showers.

Half-and-half models

Half-and-half models look very nearly like one-piece showers. The advantage to this type of sectional unit is that it has only one seam, and it is horizontal. In theory, water should run down the wall and never enter the seam. If water accumulates in the crease of the seam, an inner lip will normally repel it. For these reasons, many plumbers don't seal the seam with caulking, and there is rarely a problem. But, I have seen problems with these types of units.

About 3 years ago, one of my plumbers was working on a job with a half-and-half sectional shower. The unit was in an upstairs bathroom and it leaked randomly. When it leaked, a substantial amount of water gathered in the ceiling below and escaped through a light fixture.

The plumber on the job was a very good service plumber, but he couldn't solve the puzzle of what was leaking and why it only leaked periodically. Having worked with me on similar jobs, my plumber went through all the right troubleshooting steps, but came up empty. He called me to the job.

I talked with the plumber and the homeowner to gain as much knowledge of the problem as possible. The fact that the shower only leaked occasionally was troubling. Everyone in the family used the shower, but it didn't always leak. After gathering all the data available, I went upstairs to inspect the shower.

I suspected the shower curtain as the problem, but my plumber had already tested the floor tile, and it didn't leak. The top edge of the shower had not been caulked, but no one in the family was extremely tall, so that didn't seem to be a factor. I went through all the normal troubleshooting procedures, and the shower never leaked.

I was about to ask the homeowner to keep track of when the shower leaked in the future, and who was using it at the time, and to call us when there was more information to work with. We had looked for all the common problems and couldn't find any. Then I noticed that the shower head was a hand-held unit mounted on a bracket.

I thought for a moment and asked the homeowner about her children. It turned out she had a young son. I speculated the height of child and came up with an idea. When the homeowner told me her son used the shower head as a hand-held unit, I was almost sure I had discovered the mystery leak.

Both my plumber and the homeowner watched intensely as I removed the shower head from its bracket. I cut the water on and aimed the shower head at the horizontal seam in the two-piece sectional shower. I moved the shower head much like a small boy might do if he were imitating the actions of a fire fighter.

After several minutes of spraying the walls we all went downstairs to see if there would be any leak. Guess what? That's right, I found the leak.

I'll admit that the seams of two-piece showers probably were never meant to stand up to the torture test I had given that one, but the uncaulked seam was the culprit. My plumber caulked the seam and the shower has never leaked since.

Leaks like this can drive a plumber crazy because they shouldn't be happening. Who would ever think of a child shooting water into the seam for fun? I got lucky.

Molded Shower Bases

Molded shower bases, in themselves, don't generally cause problems. If there is a problem with a molded base, it is usually with the drain or surround walls that are too short and caulked improperly. We've already talked about the importance of caulking and where the caulking should be, so let's concentrate on the drains for shower bases.

There are two common types of drain arrangements for molded bases and shower units. One is a metal collar that is meant to be caulked with lead or filled with a rubber gasket. The other is a screw-in drain. The two vary in their ways of leaking, so we will look at each time individually.

Metal collars

The metal collars on molded bases are factory installed, and I've never seen a collar leak. I have, however, seen plenty of collar joints leak. It is amazing what some creative homeowners and inexperienced plumbers will come up with to seal the area between the drain pipe and the collar.

Most modern plumbers use special rubber gaskets to make a legitimate seal around the drain pipe. Some old-school plumbers, myself included, still use oakum and lead. Both of these methods are fine, and usually neither leaks when installed by a professional.

The first step to take when looking for a leak in this type of base is to remove the strainer from the drain. Then you should carefully inspect the joint between the collar and the drain pipe.

Normally, water runs out of the shower base and down the drain pipe quickly enough that serious undershower leaks don't occur. However, if the joint in the collar is not a proper one, heavy water damage can be done. Let me give you an example.

I responded to a call years ago where a shower had dumped so much water into the ceiling of a rental property that a portion of the ceiling actually collapsed. Apparently the leak had gone on for quite a while before the tenant notified the landlord.

With the ceiling out of the way, I could see the underside of the shower base very well. The plywood subfloor was black with water damage, and the joists had been affected. It was obvious the leak was from the drain in the base.

I went upstairs and removed the strainer from the drain. Looking inside, I saw what appeared to be a homemade seal of glue or clear caulking. After I picked the mess out of the collar I found remnants of newspaper stuffed around the pipe. Someone had tried to seal the joint with improper materials. I suspect they used newspaper to close the void and then dumped the PVC glue or caulking in around the pipe, hoping to make a watertight seal. The idea didn't hold water.

On another occasion I removed a strainer and found lead wool sealing the drain of a shower base. The problem with this job was a lack of oakum. If oakum had been packed around the pipe and then capped with lead wool, it probably wouldn't have leaked, but without the oakum, the lead seal was ineffective.

Don't expect this type of drain to be properly sealed. Most homeowners, and some plumbers, don't realize they will need a special shower gasket or oakum and lead to seal these joints. When they attempt to make the joint, they are at a loss for what to do, and people sometimes improvise.

Screw-in drains

Screw-in drains are easy for most people to install. They look so easy to install that mistakes are often made. Putty isn't put under the rim of the drain, the fiber gasket is forgotten or lost, and the lock ring isn't always tight. Any of these conditions may result in a leak. Unfortunately, you can't check for these deficiencies without gaining access to the bottom of the drain, and this frequently means cutting out a section of ceiling.

Pan-Type Showers

Pan-type showers are the worst when it comes to pinpointing leaks. There are so many places where leaks are possible that it can be all but impossible to identify the exact location of a leak. All the normal possibilities for leaks around escutcheons and wall tile exist, plus you are faced with a base that could be leaking around the drain or through the pan itself.

Depending on the age of the shower pan, it might be made from sheet lead, copper, coated paper, or membrane material. Some pan material is not installed properly, and others give out with age and the stress of settling.

Most leakage with pan-type showers is related to the drain. Either the wrong type of drain is used or the right drain is used but installed incorrectly. Many young plumbers have limited to no experience with pan showers.

I once had a plumber install a screw-in drain in pan material. This obviously was not the right thing to do. He glued the drain on the drain pipe and left it sitting above the pan material at the height expected for the final base. Fortunately, I found the inappropriate installation before the base was poured. If I hadn't, the shower would have leaked, and my company would have probably been sued for damages. If a licensed plumber can make this type of mistake, you can imagine what do-it-yourselfers might come up with for drain options.

When the flanged drains that are meant to be used with pan showers are installed properly, they rarely leak. It is the improvised drains that you must really look out for.

There is also the possibility that a pan liner was never used for a built-up shower base. I have seen this be the case a couple of times. Inexperienced people weren't aware that a pan liner was needed, and they simply poured the base on the subfloor.

Regrouting the tile on the shower floor will take care of some bad-pan problems, but it may be necessary to remove and replace the existing floor with a new one.

If the drain is leaking, repairing it requires damaging the shower base, and trying to match tiles for the repair probably won't be feasible. Before you decide to break up a portion of the base to inspect for trouble, look below the base for a less-expensive option. There is normally much less trouble and cost involved in cutting and repairing a ceiling than there is in damaging a tiled shower base.

The key here is to know where the leak is coming from before you jump in over your head by breaking into the base. It would be a shame to destroy a base only to find that the wall tile needed new grouting.

Metal Showers

Metal showers are not very common, but they do exist, and they often develop leaks. Metal showers share many of the same leak possibili-

ties that other showers do, but they have one unique way of leaking—
they can rust out. When you are troubleshooting metal showers, be
sure to look for any signs of rust or holes.

Emergency Showers

Emergency showers don't offer any particular challenge in troubleshooting. There is often no drain or base installed with an emergency station. Typically, the only troubleshooting with these units
involves the valves, and that information is covered in Chap. 10.

This concludes our work with showers. We are now ready to move
on to the troubleshooting methods used for spas and whirlpools.

7

Troubleshooting Spas and Whirlpools

How much do you know about troubleshooting spas and whirlpools? Do you get many calls for this type of work? Spas and whirlpools don't normally cause much trouble, but when they do, the problems can be difficult to diagnose properly. Unless you are something of a specialist on the subject, spas and whirlpools can make you scratch your head in short time.

Like faucets, there are so many different makes and models of spas and whirlpools that keeping up with all of them is nearly impossible. There are, however, many characteristics that are similar in all the various brands of both spas and whirlpools. It is that common thread that we are going to approach in this chapter.

To look at a whirlpool that is already installed, there doesn't seem to be a lot that could go wrong with it. There's the tub filler, the drain, and a few jets in the sides of the tub. Pretty simple, right? Wrong. Whirlpools have pumps, controls, suction fittings, O-rings, air lines, air-control valves, and other parts that can cause problems for the plumber in the field.

Plumbers who mostly do new work don't often have to figure out why the jets on a whirlpool are not making the water in the tub swirl. Typically, these plumbers just install the tubs and forget about them. But what happens if a new-construction plumber installs a whirlpool and the unit fails to operate properly? Who is going to troubleshoot the job and correct the problem?

If the failure occurs within the first year of installation, the plumbing company that installed the whirlpool is probably going to have to make the repairs under warranty service. Under these conditions the company has only a few options. Either a plumber from the company will have to know what to look for, where to look for it, and how to fix the problem, or the company may have to call a factory representative. The only other option will be to call in a plumber from a different company who has the ability to repair the whirlpool.

Plumbers who make their livings doing service and repair work will get calls from time to time that require them to work with spas and whirlpools. Even though plumbers are not needed to install many residential spas, they are called to correct deficiencies when the spas break down.

Do you know how to remove the apron from a spa? Does a spa have a ground-fault intercepter (GFI) breaker? Are there any strainers that may get clogged up on a spa? You may have answered all of these questions correctly, but many plumbers have never had to remove the apron from a spa. And, I'm sure a number of professional plumbers have never done any type of repair work on spas.

Whirlpools are more common in most homes than spas, but spas are becoming more popular. The fact that they can be installed without the help of a plumber, they don't have to be filled with water before every use, and the water is always heated and ready for use makes them a great alternative to large whirlpool tubs. As the trend for portable spas continues to grow, the demand for plumbers who know how to work on them will grow.

Not only are plumbers often required to know how to fix spas and whirlpools, they are frequently looked to for answers to common questions. For example, do you know what the maximum recommended temperature of water in a spa is? How long can an adult stay in the hot water of a spa safely, under normal conditions? If you can't answer these questions, your customers may doubt your ability to install or maintain their spas.

Did you know the answers to the two spa questions? Well, if you didn't, the maximum recommended temperature is 104 degrees F. It is normally accepted that healthy, normal adults should not stay in the water for more than 15 minutes at a time. You'll get more tips like this as we go along.

Spas

We will begin our look at troubleshooting spas and whirlpools with a study of spas. First we will look at the various parts and aspects of spas that are most likely to affect you on a daily basis. Then, we will delve into the problems and remedies that you may work with on spas. Let's start with a look at the major components.

Pumps

The pumps in spas are what make the water move. They do this by pulling the water in at one end and pushing it out at the other. The water coming from the pump is under great pressure as it is forced through the piping that delivers it to the outlets in the spa. The suction fitting in the spa allows the pump to pull the water back in and recirculate it.

When the pump pulls water in from the suction fitting, it directs it to a filter and then to a heater. During this process, clogs can occur in the strainer basket on the pump. Once the water has been filtered and heated, it is pumped back into the spa through the hydrojets.

Spa pumps should be sized to match the filter's capacity to trap oils and organic particles. If water is not circulated through the filter at the proper flow rate, the filter cannot do its job effectively.

Additionally, the pump should be rated in horsepower so that it is sized properly for the volume of water required by the spa. The number of hydrojets will also affect the sizing of the pump. You see, it takes a lot more effort for the pump to force water through the jets than it does to simply circulate water. As an example, a pump with a rating of 1 horsepower should be capable of handling the requirements of a spa with a capacity of up to 700 gallons of water and four hydrojets. This sizing is not carved in stone, but it is a benchmark to work from.

Many high-quality spas have two-speed pumps. These pumps are more energy efficient and are favored for their cost-saving performance. The low-speed setting on the pump is used to circulate water when the jets are not being used, and the high-speed setting is used to activate the hydrojets.

If you get into working on large spas, it will not be unusual to come across pumps rated up to 2 horsepower. It also would not be uncommon to find a two-pump system on large spas. There may be a small pump that handles the average water circulation and a larger pump that produces the force to run the hydrojets.

Filters and skimmers

Filters and skimmers are key elements in the makeup of spas. These devices not only keep the water clean, but they prolong the life of the spa. Without such devices, the spa water could become contaminated with algae, dirt, sand, and other unwanted materials. This type of contamination could be detrimental to a clean bathing experience, and it could cause problems with the pump and plumbing.

Filters that run on 2-hour cycles are common. In other words, the water gets filtered for at least 2 hours each day. Skimmers are not found on all spas, but they are a desirable feature. The skimmers pick up floating debris from the spa water and deposit it in a removable, cleanable basket.

Sometimes the skimmer is combined with the filter in a very convenient top-loading design. This allows the unit to be cleaned without draining the spa. On older, and less-expensive, models it is sometimes necessary to empty the contents of the spa to access the filters. Typically, cartridge filters can operate for up to 2 years without being cleaned.

Diatomaceous earth filters. Diatomaceous earth (DE) filters are used on large spas, such as you might find in a health club. These filters are not so easy to maintain. The makeup of a DE filter allows it to trap the much larger quantities of dirt and unwanted ingredients that may be found in the water of a spa.

Unlike cartridge filters, which can be cleaned with a garden hose, DE filters require a backwash cycle for cleaning. Once the backwash is complete, a new coating of DE should be applied.

Sand filters. Sand filters are also used on large, commercial-grade spas. Like DE filters, these high-rate sand filters require a backwash cycle when they need to be cleaned.

Heaters

Heaters are essential to the favorable operation of a spa. The heater must be sized to meet the demands of use placed on the spa. It is possible that a spa may be equipped with either a gas or electric heater. Let's talk about each type of heater for a moment.

Gas heaters. Gas heaters are desirable for their ability to heat water quickly and cost effectively. However, the need for gas piping and venting can be enough to turn customers to electric heaters.

When you are asked to work on a spa with a gas heater, you will be dealing with the following heater components: a water inlet, a water outlet, a combustion chamber, a burner, a pilot light or electronic ignition, a heat exchanger, heater controls, a vent, and gas piping.

Electric heaters. Electric heaters for spas come in both 110- and 220-volt ratings. Most residential spas can be fitted with the 110-volt versions. There are also electric heaters that are convertible to either 110 or 220 volts.

There are some drawbacks to 110-volt heaters that can affect the troubleshooter in the field. Electric heaters for small spas are small themselves. This limits their ability to heat water quickly. In fact, it is common for these heaters to be rated to produce only 2 degrees of rise in water temperature for each hour they are in use. This is bad enough, but there is more.

Not only are electric spa heaters slow, they pull a lot of electricity, usually the full amount that a 20-amp household circuit can handle safely. Therefore, the hydrojets and air blower on the spa are unable to run at the same time the heater is in use. If the spa is being used frequently, the heater may not be able to keep up with the demand for hot water. You will do well to remember this fact. This type of problem, however, is not likely to occur with electric heaters rated at 220 volts.

There is at least one brand of spa that manages to heat its water with the use of its pump. The dealer I spoke to on this particular model said that the cost for electricity to maintain a temperature of 104 degrees F in Maine, with an indoor installation, was about $6 per month. The cost doubled when I inquired about an outside installation, but it takes hardy souls to have outside spas in Maine during the winter.

Controls

Every spa you work with will be equipped with controls. Some of these controls may be operated manually, and some of them may work automatically. The controls used to start the pump or blower are normally mounted on the rim of the spa. They are sometimes mounted in the top surface of a deck that surrounds a spa. They may be touch-activated electronic switches or air switches. In the case of air switches, the act of depressing them sends air down a piece of tubing to activate the equipment switch.

Controls for the heaters used with spas are normally automatic. They are thermostatically controlled. Some spa heaters, typically those rated at 220 volts or that are gas fired, have time-clock controls. These more efficient heaters don't require as much running time to heat water, so their cycle can be set to coincide with the usage of the spa. A few sophisticated models even have a two-clock system that allows more flexibility in the time settings for heater activation.

Remote controls are also available on some types of spas. These controls are convenient for users who have spas outside of their homes. With the remote control pad, the user can program the spa without stepping outside the home.

Blowers

Many spas are equipped with air blowers. These devices produce the gushing flow of bubbles that is found in spas. While the air-induction systems don't normally require much maintenance, they can hinder the spa's ability to maintain a low heat loss. The blowers can also be a nuisance in terms of noise.

Air blowers for most spas pull in ambient air. If the air being blown into the spa is cold, such as may be the case in an outdoor setting, the air can cool the water more quickly than a small heater can compensate for the temperature changes.

Blowers run at high speeds and can produce annoying sounds. If the blower whines too loudly, it sort of ruins the atmosphere of a nice, relaxing soak in the spa. If your customer complains that a spa is too noisy, the air blower is the first place to turn your attention to.

The jets

The jets on spas are what make the warm water magical. They are also the components that most users notice problems with. Correcting the problems is often as simple as setting the air intake to a more appropriate setting.

Average hydrojets push out water at a rate of about 15 gallons per minute, but some spas are set up to allow all but one jet to be closed, producing a push from that single jet in the range of 90 gallons per minute.

Now that you are aware of the basic components that make up a spa, let's examine some of the routine maintenance responsibilities that go along with long, hot soaks in the spa.

Water Maintenance

The water maintenance in spas is not normally a plumber's responsibility; it is the job of the spa's owner. However, a lack of routine maintenance on the part of the spa owner can have an affect on a plumber called in to troubleshoot problems with the spa. Let's look at how water maintenance may affect you as a service plumber.

Potential hydrogen

Potential hydrogen, or pH as most plumbers refer to it, can cause trouble with spas. If the pH is too low, below 7 on the pH scale, parts of the spa equipment can be damaged by the acidic water. The lower the pH is, the more likely this is to occur. A low pH can also result in complaints from the users of the spa. Their eyes and skin may sting, burn, or show other ill effects from highly acidic water.

At the other end of the spectrum, a high pH count, something over 8, can cause the spa water to become cloudy. It can also produce scaling on the spa equipment, and mineral deposits can cause problems with the plumbing. The heaters of spas are particularly suspectable to this type of problem. For example, if the heater coils in an electric heater become encrusted with mineral deposits, the heater can require substantially more time and energy to maintain the water temperature in the spa. This is similar to what happens with the heating elements in electric water heaters for homes and businesses.

The water in a spa should be tested regularly. If problems are found, chemicals can be added, like soda ash, to correct the problems before they escalate. The most effective way to avoid mechanical and equipment problems with spas is to maintain the water at chemical levels recommended by the manufacturers.

Troubleshooting Spas

Troubleshooting spas should not be a job that intimidates you. Even if you are unfamiliar with the operation and principles of spas, don't despair. The learning curve on spas is short. To prove this point, let's examine some common complaints that come into plumber's shops on spas.

My spa is full of foam

How often have you been called by a customer who said, "My spa is full of foam"? Maybe you have never gotten such a call, but these calls are common for plumbers who are known to work with spas.

When a customer's spa is foaming at the mouth, so to speak, the customer can be more than a little panicked. They may believe that there filter has died or that their pump is malfunctioning. In all cases, they will want you, their trusted plumber, to solve their problem. The problem is, many regular plumbers have no idea what is causing the foaming action. Do you know what the cause of the problem is?

When the water in a spa begins to foam, it is an indication of poor usage of the spa. It is not that any of the equipment is failing; it is an operator error. The water being pushed out of the hydrojets stir up the water in the spa. If shampoo, body lotions, or in some cases cosmetics have gotten into the spa water, the constant agitation from the jets will whip the water and its contents into a foaming monster.

I once had a customer call me late at night who said that her spa was foaming up over its overflow rim. She was terrified that the foam would consume her entire exercise room. I knew that the problem was not serious, but she didn't.

It seems that the woman's daughter had gotten out of school early on this particular day and had spent considerable time in the spa. During her soak she washed her hair, not a smart thing to do in a spa. The left-over shampoo was what caused the massive foaming.

If you have a customer who is frightful of foaming, suggest an antifoaming compound. There are many types available that will control the natural foaming of spas. In extreme cases, like the one I was called to, you may have to skim off the foam before the antifoam agents can take over and win the battle.

Cloudy water

Complaints about cloudy water in spas are not at all uncommon. There are a few possible causes for the cloudy water. The first logical troubleshooting step to take is an inspection of the filtration system. If the filter is set up to run at least 2 hours a day, for most residential applications, and if the filter is clean, you must look elsewhere.

Test the water for its pH rating. A high pH number, something above 7.6, can result in cloudy water. If the pH is the problem, it can be treated with chemicals, like soda ash, to control the problem and to clear up the water.

A third possible cause could be a lot of undissolved solids or an overabundance of chemicals. When this is the case, you should drain the spa and refill it with fresh water.

Hard water

Hard water can produce scaling problems with spas. The effects can come in the form of hard deposits on the sides of the spa, in the filter, or in the heater (this is one of the most common locations). The scaling is a result of an unbalanced pH rating and hard water. Appropriate chemicals introduced into the spa water will solve the problem.

Burning eyes

Some customers may complain that they have burning eyes after using their spas. This type of problem can occur when the pH rating is out of balance (a pH number that does not fall within the range of 7.4 to 7.6). It is also possible that chlorine is causing the burning sensation.

Chlorine can combine with nitrogen that is present from perspiration or cosmetics to form chloramines. The odor from this concentration can irritate human eyes. Chlorine is not the only culprit that is capable of producing irritating gases. Bromine, which is used to clean spa water, can produce bromamines. Unlike chlorine, the production of bromamines is odorless, but it can be equally irritating to the eyes.

If the pH rating for the water tests is okay, the best alternative is to drain the spa and refill it with fresh water.

Algae

Algae will build up in a spa if the unit is not covered between uses and if the water is not treated properly. When algae is a problem, chlorination or draining and refilling the spa is the logical solution.

Ugly water

Ugly water is not something that customers want in their spas. When spa water is discolored or stains the fixture, it is a sure sign that some type of metallic content is in the water. There are chemicals available to help in the removal of metals in the water, but draining and refilling the spa may be all that is necessary.

What happened to my bubbles?

If you work with spas, sooner or later you will get a call from a customer who asks, "What happened to my bubbles?" A pretty good guess on this problem is that the jets in the sides of the spa have been closed down. Kids seem to be fascinated with their abilities to turn the rings and cut off the bubbles. Parents who don't know their children have been playing with the hydrojets often don't think to check the settings before calling in professional plumbers.

It is, however, possible that the pump has lost its ability to flush water through the jets. Therefore, if the jets are open, check out the pump. If you have reason to suspect the pump, check the strainer basket. It may be that the strainer is clogged and is prohibiting the pump from picking up water from the suction outlets that can later be pumped through the jets.

The filter could, in extreme cases, be at fault for puny pressure at the jets. If the filter is clogged, the water cannot circulate properly. Under these conditions, clean the filter and see if things get back to normal.

Don't be fooled by a two-speed pump. Remember that the pump only circulates water at its lower speed, and it needs the high-speed function to kick in before it can activate the jets. In the case of dual-pump systems, you must troubleshoot both pumps, but concentrate on the larger pump first.

Tepid water

Tepid water in a spa is caused by a failing or inadequate heater. It may be that an air blower is pulling cold air into the water more quickly than the heater can compensate for the temperature change. The heater may be too small for the spa, but this is unlikely, unless the factory-installed heater was replaced.

Constant or frequent use of a spa often causes the water temperature to drop when the heater is an electric 110-volt unit. Since these heaters don't produce a lot of hot water to begin with, and they can't run while the jets are in operation, it is easy to overpower them to the point of turning the water into a lukewarm bath rather than a hot, relaxing soak.

Another possibility is a failure of the heater's thermostat. If the thermostat is bad, the heater cannot heat the water properly. In larger units, the time-clock settings may be out of adjustment. We will address this issue in just a few minutes.

The pump won't run

When you are dealing with a spa where the pump won't run, check the circuit breakers or fuses. Most spas are equipped with GFI devices for safety purposes. Check the GFI and the panel box to see that everything is in good order. And, don't forget to see that the spa, if it is a portable model, is plugged in to a suitable electrical outlet. If all the simple electrical work is intact, check the windings on the pump and the controls.

Clogged filters

Clogged filters on spas are a common complaint. In the case of cartridge filters, you can simply replace the filter insert. For sand and DE filters, you will have to backwash them.

Time-clock problems

Time-clock problems are often associated with spas that are set up with gas-fired heaters or heaters with 220-volt ratings. If the time-clock settings get out of adjustment, the heaters on these spas will not warm the water when the customer wants it to. This results in a cool dip, instead of a hot soak.

We have covered the bases for troubleshooting spas pretty thoroughly, so now let's concentrate on the issues pertaining to whirlpools. While some of the situations are similar between whirlpools and spas, there are many differences to consider.

What Makes a Whirlpool Different?

One of the biggest differences between a whirlpool and a spa is that whirlpool tubs are not designed to hold their water for multiple uses. Where a spa is always ready for use, whirlpools are meant to be filled with water before each use. This is a sizable difference, both in time, convenience, and cost.

The time it takes to fill a two-person whirlpool is considerable. The fact that whirlpools have to be filled before they can be used is not conducive to frequent use. It often takes more time to fill a large tub with water than people are willing to spend. In fact, it is not unusual for the filling time to exceed the soaking time.

Some large whirlpools have on-site water heaters, but many of them are filled from a large water heater that serves the rest of the plumbing fixtures. The filling of a large whirlpool not only depletes the supply of hot water quickly, but it is a costly proposition. Heating the volume of water needed for these tubs requires a lot of energy. Even the mere cost of the water used to fill the tubs can add up over the course of time. For all of these reasons, spas are gaining more attention.

I have owned large whirlpool tubs, a soaking tub, and a spa. Of all these units, the spa was always the most convenient and the most enjoyable. There is no doubt in my mind why the buying public is so pleased with the overall performance and characteristics of spas. However, as fashionable and practical as spas are, whirlpools continue to command their share of the marketplace.

Some of the other differences between spas and whirlpool tubs are not as obvious. For example, a whirlpool tub does not come equipped with a skimmer, but then, it doesn't need one. And as I mentioned earlier, many whirlpools don't have independent heaters; this is another difference between whirlpools and spas.

Since whirlpools don't retain their water for multiple uses, there is no need for the chemistry tests and treatments that are mandatory with spas.

Time clocks are not needed on whirlpools. Since the tub doesn't keep its water on board, there is no need to set a timer to turn on a heater to warm the water.

The Similarities between Whirlpools and Spas

Whirlpools and spas both rely on pumps to circulate water and to produce force at the hydrojets. Both units are fitted with hydrojets that bubble the water, and both types of tubs have suction intakes. Another similarity is the shared use of air-control devices.

Troubleshooting Whirlpools

Troubleshooting whirlpools is not much different from working with a regular bathtub. The tub filler is similar to that of a regular bathtub, and the waste and overflow is almost identical, except that it usually has a higher overflow tube.

The differences between whirlpool tubs and standard bathtubs come into play when the problems being experienced have to do with the whirlpool pump, fittings, piping, jets, and related gear (Figs. 7.1, 7.2, and 7.3).

The suction fitting on a whirlpool can become clogged and need cleaning. This doesn't happen often, but it is something you should be aware of. If the jets are not operating properly, check the strainer in the suction fitting.

If the air-volume control is not adjusted properly, a good flow of water is not going to explode out of the hydrojets. When this is the case, a few quick turns of the air-volume knob will normally solve the problem.

If the eyeball jets in the tub are giving you trouble, you can purchase a repair/replacement kit to remedy the situation.

There are O-rings at both the suction and discharge connections of

EMPRESS
Acrylic Whirlpool Recess Bath

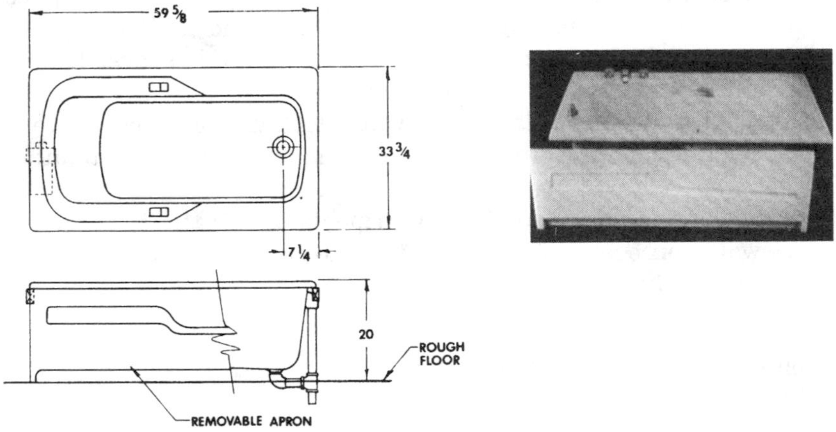

Dimensions are Nominal

Bath: 2-925W *Empress* acrylic whirlpool bath with optional 2-297 *Apron* for recess installation, four 360° adjustable whirlpool jets, variable air-injection control, 0-30 minute remote timer and ¾ HP whirlpool pump. Sloping back, with grab handle and integral arm rest on both sides.

Trim:

Waste:

Color:

Figure 7.1 Whirlpool tub with removable apron. (*Courtesy of Crane Plumbing.*)

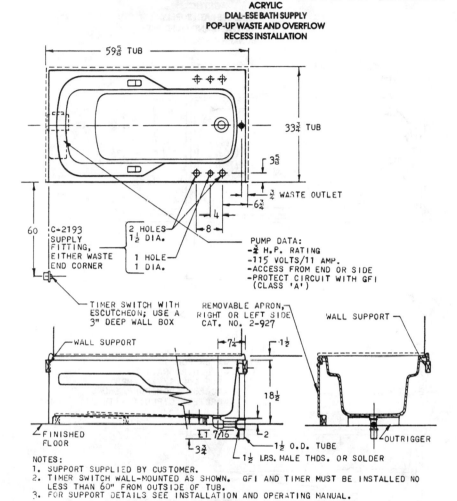

Figure 7.2 Rough-in details and description of a whirlpool tub. (*Courtesy of Crane Plumbing.*)

the whirlpool pumps. The O-rings don't normally act up, but if they do, they are easy to swap out.

The pump and control box for a whirlpool is not unlike those for other plumbing applications. You can use a standard test meter to troubleshoot the pump and controls.

Many whirlpools are equipped with personal shower attachments. These devices require a backflow preventor, but this part of the fixture shouldn't give you any trouble.

ATLANTIS WHIRLPOOL BATH
ACRYLIC
DIAL-ESE BATH SUPPLY
POP-UP WASTE AND OVERFLOW
SUNKEN INSTALLATION

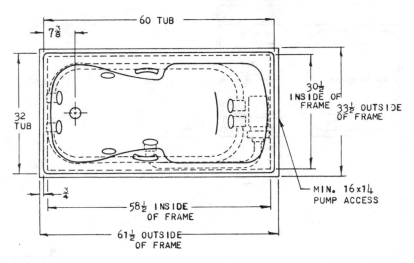

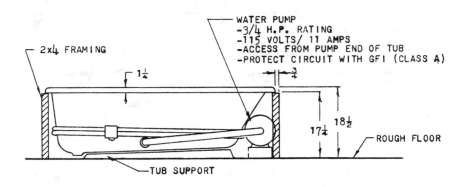

NOTES:
1. FRAME AND GROUND FAULT INTERRUPTER (G.F.I.) SUPPLIED BY CUSTOMER.
2. FOR SUPPORT DETAILS, SEE INSTALLATION INSTRUCTIONS.

Figure 7.3 Rough-in details and description of a sunken whirlpool tub. (*Courtesy of Crane Plumbing.*)

Hard water can take its toll on the mechanisms of a whirlpool, but really the consequences are no worse than they are for any normal plumbing fixture.

Usually, the most difficult and frustrating part about working on whirlpools is the limited access provided for service and repairs. The pump and controls are usually placed near the apron of the tub, but some of the fittings and tubing can be nearly impossible to see, reach, or work on.

The Basics

The basics of troubleshooting whirlpools and spas are not that different from those used on any other plumbing problems. The key is to have a good working knowledge of the brand of fixture you are working with and to move about your troubleshooting in a structured manner.

Manufacturers of the various tubs will generally supply you with detailed cut sheets on their products if you request them. Some of the companies will even throw in a few troubleshooting tips. If you order this type of information, you can do a fine job of troubleshooting both whirlpools and spas.

The next chapter is all about troubleshooting bidets. Like spas, bidets are not one of the more common fixtures for plumbers to work with. Since bidets are not an everyday household fixture, you will probably learn a lot when you turn the page.

8

Troubleshooting Bidets

Troubleshooting bidets is not a job most plumbers do on a daily basis. In fact, some plumbers have never worked with bidets. This doesn't mean, however, that you will never be asked to service or repair one.

Bidets are not a common plumbing fixture in most American homes, but they are prevalent in many upscale housing developments. At first glance, bidets don't appear too intimidating (Figs. 8.1 and 8.2), but a number of plumbers have no idea of how they work or how to troubleshoot them.

This chapter is going to put bidets in perspective for you. We will discuss various types of bidets, their drainage systems, and their valves. Once you've finished this chapter, you will know the ins and outs of bidet plumbing. Let's start this section with a question-and-answer format to see how much you already know about bidets.

1. *Do bidets have integral traps?*   Toilets have integral traps, so it would be easy to assume bidets do also, but they don't. Traps for bidets are installed below floor level as are bathtub traps.

Trap leaks under bidets can do a lot of damage, and they can be hard to find. If the slip-nut washers allow water to leak out of the trap or drainage tubing, the water can collect for a long time before it is noticed.

When a bidet is located on a second floor, the first evidence of trap leaks is usually a stained ceiling. The stain, however, will not always be found directly beneath the bidet. It is not unusual for the water from the leaks to run across the ceiling, follow electrical wires, or seek an opening, like a light fixture, to run out of.

If the water just lays on a ceiling, it can take days, or even weeks, for it to work its way through it. Small leaks can take months to show up. When these leaks are found by homeowners, severe damage has

BIARRITZ
Vitreous China Bidet

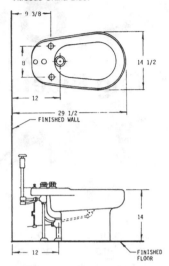

Biarritz Bidet 3-277

Bidet:	3-277 **Biarritz** vitreous china bidet with rose-spray, integral overflow and flushing rim with shelf. extended back shelf.
Trim:	C-3046 chromium plated two-valve supply fitting with backflow preventer, transfer valve and rose spray; indexed acrylic handles and pop-up waste with 1¼" O.D. tailpiece.
Trap:	
Supply:	
Color:	

Sectional view of
Biarritz bidet, showing
assembly of the built-in
spray.

Figure 8.1   Bidet. (*Courtesy of Crane Plumbing.*)

usually already occurred to the ceiling. The ceiling joist may even be soaked before the leak is discovered.

Finding a hidden ceiling leak that is the result of a bidet trap is similar to tracing leaks from bathtubs and showers. The key to finding the leak in the trap is filling the bowl of the bidet to capacity and releasing all of the water at once. I've said it before, but I'll say it again. Too many plumbers fail to fill bowls to test drainage systems, and without the pressure provided from a full discharge of water, many leaks will not show themselves.

2. *Do bidets use both hot and cold water?* Yes, they do. Bidets are equipped with valves that allow the hot and cold water to be mixed to a suitable temperature.

BIARRITZ BIDET
VITREOUS CHINA
TWO VALVE SUPPLY
BACKFLOW PREVENTER
TRANSFER VALVE & POP-UP WASTE

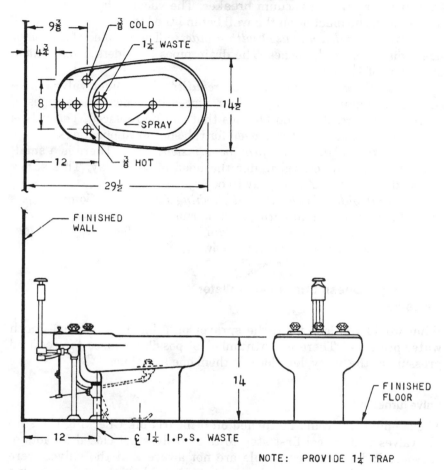

NOTE: Installation instructions and mounting bracket included with bidet.

Figure 8.2 Rough-in details and description of a bidet. (*Courtesy of Crane Plumbing.*)

3. *Do bidets have flush handles?* No. Unlike toilets where a water reserve is flushed from a tank or flush valve to clean the bowl, bidets work on a principle similar to lavatories. They are equipped with pop-up assemblies and lift rods that allow them to hold and drain water.

4. *Do all bidets have sprayers in their bowls?* No. Some bidets use over-the-rim faucets and do not have sprayers.

5. *Are vacuum beakers required on bidets?*   Bidets that use over-the-rim faucets and that are not equipped with sprayers are not required to be fitted with vacuum breakers.

Bidets that are equipped with sprayers are required to be protected from backflow with a vacuum breaker. The vacuum breaker may be mounted on the bidet or on the wall behind the bidet.

6. *What size drain is used with a bidet?*   The tailpiece for a bidet has a diameter of $1\frac{1}{4}$ inches. The drain pipes for bidets usually have diameters of $1\frac{1}{2}$ inches.

7. *Can the valve for a bidet be rebuilt?*   Yes. Bidet valves have seats and diaphragms that can be replaced if needed. Some types of valves have cartridges and O-rings that can be replaced. The valves for bidets are similar to those used for tub/shower applications.

8. *Can the height of the spray be adjusted?*   Yes. There is a small adjustment screw located under the head of the spray. This screw allows the intensity of the spray to be adjusted.

9. *Do all bidets have integral flushing rims?*   No. Some models are not equipped with integral flushing rims.

10. *What does the transfer valve do?*   The transfer valve for a bidet diverts water to the sprayer in the bowl.

The Sprayer Doesn't Have Much Water Pressure

What would you look for if the sprayer on a bidet didn't have much water pressure? There are a number of possible causes for reduced pressure at the spray; let's look at them one at a time.

Valve turned off

The supply valve could be turned off or nearly off. Checking the supply valves is a logical first step to take. Valves get closed for lots of reasons, and sometimes people are not aware that the valves were tampered with. For example, a curious child may turn the valve toward the closed position. If this were to happen, the parents might have no idea the valve has been turned.

When you are troubleshooting, you cannot afford to overlook the obvious. I know the idea of having a valve closed inadvertently may be a long shot, but I have found this to be the case on toilets, so it could certainly happen with a bidet.

Clogged sprayer

A clogged sprayer could result in low water pressure and a lackluster fountain of water. Houses with private water supplies often develop problems with the buildup of minerals, and this could cause the flow of a bidet sprayer to be minimized.

In addition to mineral deposits, other types of debris could plug up the sprayers. Sand, gravel, loose pieces of copper shavings, and similar materials could cause the spray holes to become clogged.

Crimped hose

A crimped hose can definitely slow down the flow of water to a bidet. While it is not likely that a hose will become crimped or pinched by itself, it is worth checking. Normally, the lack of water will not be caused by a crimped hose unless the bidet was recently installed or worked on.

Obstructions

Obstructions in the water supply can slow down the delivery of water, and it doesn't take a very large object to restrict the flow of a bidet supply. You can check for obstructions in the tubing by taking sections apart and inspecting them.

Transfer valve

If the inner workings of a transfer valve goes bad, getting water to the sprayer will not be easy at all. You can think of the transfer valve as being like the diverter handle on a three-handle tub/shower valve. If it's not in good working order, the water cannot be diverted properly.

Debris in the supply valve

Debris in the supply valve is another possible cause for reduced water pressure. This situation is not common, but it can occur. If gravel, sand, or a copper shaving has become lodged in the supply valve, you will have to disassemble the valve to clear the obstruction.

Restricted supply tube

Some bidet faucets use standard supply tubes, and a restricted supply tube can affect the water pressure at the faucet. The supply tube could be crimped, or it could have an obstruction inside of it. This is a simple problem to eliminate; just remove the supply tube and inspect it. Replace the tube if necessary.

My experiences

My experiences with reduced water pressure under conditions similar to the ones described above have been numerous. To put the troubleshooting into real-life terms, let me share a few of those experiences with you.

Bread in the supply tube.   I was once called to troubleshoot the reason there was no water pressure at a faucet. I went through about the same troubleshooting procedures I have explained to you. The only step I left out initially was checking the supply tube for an obstruction.

After checking the cutoff valves, the supply hoses, the diverter valve, and everything else I could think of at the time, I was stumped. Only the cold water side of the faucet was being affected, and no other fixtures in the home were suffering similar symptoms. Needless to say, I was confused and frustrated.

I was so sure I had checked everything that could be causing the problem that I went back through each phase of the troubleshooting again. The results were the same; I found nothing that would account for the low water pressure.

The house I was working in was relatively new, so I ruled out a rust closure in a galvanized supply pipe. Not wanting to give up, I sat down on the floor and stared at the fixture, trying to think of something else to try. My meditation didn't produce any new ideas. Baffled, I went downstairs to talk to the homeowner.

After questioning the homeowner on how long the problem had existed, I was surprised to find out that it had started shortly after another plumber had been to the house to repair a water line that had frozen and burst in the outside wall, below the bathroom.

With this new information, I returned to the bathroom and promptly removed the supply tube. When I looked in the tubing, I was shocked that I couldn't see any light through it. The supply tube was plugged completely with something. It was then that I used a piece of wire to rod out the supply tube. What do you think I found?

If you guessed bread, you're right. Apparently, the plumber who fixed the broken pipe had used bread to hold back water in order to make the solder joint. When the water was turned back on, the bread was forced up the pipe and lodged in the supply tube. This was an unusual occurrence, but it was not the first time I had experienced such a situation.

As a young plumber, I once used too much bread to hold water back in a pipe I was trying to solder. The water was coming with a vengeance into the joint I wanted to solder, and I packed bread into the pipe with the end of a pencil. After I repaired the broken pipe, I was unable to get water to the toilet at the end of the supply line. Ultimately, I found that the bread I had packed in so tightly was not dissolving to let the water get to the toilet. If I had not had that experience early in my career, I might not have thought to check the supply tube in my troubleshooting of the bidet. My point is this: You cannot afford to overlook any possibility when you are troubleshooting.

Copper in the supply valve. On another occasion, I found slivers of copper tubing trapped in the supply valve to a bidet. The copper was evidently laying inside the tubing when it was installed, and the water pressure forced the copper into the supply valve, reducing the flow of water. Once I disassembled the valve and removed the copper, the problem was solved. I'll give you more of my personal experiences as we move on.

The Stains on Her Ceiling

I once had a homeowner call me to determine what was making the water stains on her ceiling. I arrived at the home and saw where the white ceiling had been discolored by water. The woman told me that she had never seen any water dripping, but that the stains just kept getting bigger. My first thought was a toilet or tub leak.

I went into the bathroom and found a shower, a whirlpool tub, two lavatories, a toilet, and a bidet. I started my troubleshooting by flushing the toilet and looking for leaks around the base; there were none. There was no condensation to be found, and all of the water supplies were dry. This led me to believe the problem was with either the tub or the shower.

This particular job turned out to be one that I am not particularly proud of. I spent a considerable amount of time trying to find the source of the water stains without success. After doing all of the routine tests on the lavatories, the whirlpool, the toilet, and the shower, I was puzzled.

I went to the bidet and turned the water on. It seemed to be working fine, and I didn't locate any leaks. At this point, I went downstairs and took some measurements to determine what fixture the stains were closest to. They were directly under the bidet. This, of course, redirected my attention to the bidet.

After wiping all of the water and drainage joints I could reach with toilet tissue, I still hadn't found the leak. Then it dawned on me. I closed the pop-up on the bidet and allowed it to fill with water. When I opened the drain, the water left slowly. I looked back under the fixture and saw the cause of the problem.

The drain for the bidet was partially blocked, causing water to drain slowly. The combination of the slow-draining fixture and a bad slip-nut washer was allowing the piping to leak where it entered the trap adapter.

I snaked the drain and replaced the worn washer. After testing the fixture again, I was confident it was fixed and that the leaking was over. To this day, I have not had any further complaints of leaks staining that woman's ceiling.

Putty under the Drain Flange

If you run across a bidet that doesn't have putty under the drain flange, you will have put your finger on a hard-to-find leak. We talked in Chap. 4 about how lavatories could empty their bowls without leaving a trace of the leak. Well, the same is true of bidets. You can apply practically all of the troubleshooting information given for drainage systems of lavatories to bidets. Let's do a rundown of all the places a bidet's drainage system may leak.

The trap

The traps for bidets can be difficult to see well. Since they are below the floor level, they can be troublesome to get to if there is a ceiling below them. It is unusual for a plastic trap to just up and start leaking, but metal traps can deteriorate with time.

If you have evidence of a leak coming from a bidet but can't find it, check the trap. Remember to fill the bowl of the bidet to capacity when checking the drainage. Just as we discussed in Chap. 4, the added pressure from a volume of water in the drainage piping can expose leaks that you might not otherwise find.

The trap adapter

You already heard about my experience with a leaking trap adapter, but don't forget what I told you. The slip-nut connection between the trap adapter and the tailpiece is not a common place to locate a leak, but it can be the source of your problem. This type of leak will generally avoid detection unless you fill the bowl to capacity and release all of the water at one time.

The threaded tailpiece

The threaded tailpiece of a bidet can leak if it is under pressure and not properly sealed. If the tailpiece is not tight, or if pipe sealant was not put on the threads, a leak can occur. It is also possible that the weakened, threaded portion of the tailpiece will corrode and create a hole. Again, this is not common, but it is usually the uncommon problems that are hardest to find.

The pivot rod

The pivot rod in the pop-up assembly provides some risk of a leak. If the retainer nut is not tight, water can easily escape around the pivot rod.

The gasket

The gasket under the drain flange can go bad and allow the fixture to leak. It is also conceivable that the nut applying pressure to the gasket may be loose, causing a leak.

The pop-up plug

The seal on the pop-up plug can be defective. When this is the case, the fixture will not hold water. You can test for this by removing the pop-up plug and inserting a rubber stopper. If the fixture still loses its water, investigate the drain flange. When the rubber stopper holds water, replace the pop-up seal or the entire pop-up plug, and your problem should be solved.

It is also possible that the pop-up plug only needs adjustment. If the lift rod is not positioned properly, the pop-up will either be held too high above the drain or not high enough. If the pop-up is too high, water will leak past it. When the pop-up is not high enough, water will drain out of the fixture slowly.

Spray assembly drain flanges

Spray assembly drain flanges that are not properly installed can generate leaks. If this part of a bidet is leaking, the cause is probably a lack of putty under the drain flange. You should loosen the mounting nut and check the installation to see that it is properly sealed.

We have just covered all the locations in the drainage system of a bidet that may leak. Now let's move into the water supply and see where leaks may exist.

Water Leaks

Water leaks can come from a number of places when working with bidets. Most of the time the leak will involve the hoses or tubing that connects the various parts of the fixture's valve and outlets. The possibilities for water leaks vary, depending upon the type of valve or faucet being used. Let's go down the list of places where you might expect to find water leaks, but keep in mind that not all bidet faucets are the same and that while some will have normal supply tubes delivering water to the faucet, others will have full-size piping feeding the valve.

Cutoff valves

As with any other type of fixture that uses standard cutoff valves, the valves for bidets can leak. The leak can occur on either side of the cutoff valve or at the packing nut. This type of leak is a pressure leak that is constant and relatively easy to find. A close visual inspection and a rubdown with toilet tissue will reveal these leaks.

Cutoffs that are screwed onto threaded nipples can leak at the threads. This can be caused by a lack of pipe dope, a loose connection, or threads that have worn through and created a hole. This situation can be fixed by removing the stop valve and making the necessary repair. The repair might be as simple as applying pipe dope and reinstalling the valve, or you might have to replace the nipple and the stop valve. Of course, the water supply to the valve must be cut off before this type of work is done.

Compression valves sometimes develop leaks from vibrations they receive during use. These leaks are normally easy to find with a visual inspection or a quick rubdown with toilet tissue. Once the leak is identified, tightening the compression nut normally stops it.

Cutoff valves that are soldered onto water distribution pipes don't generally develop leaks after having been installed for a while. It is possible, however, that water hammer could stress the joint and create a leak. Under these circumstances, the valve normally must be removed and replaced.

When stop valves are leaking around the stem of their handle, and they do frequently, all that is required to stop the leak, in most cases, is to tighten the packing nut. If the valve is extremely old, it may be necessary to remove the stem and rebuild it or replace the valve.

The compression nut that secures a supply tube in the stop valve is another common place for leaks to develop. This can happen if the supply tube is hit during routine house cleaning or if the valve is vibrated from a water hammer. These leaks can almost always be stopped simply by tightening the supply tube nut.

Supply tubes

Supply tubes can leak at either end. You've just been told what to do if the tube is leaking at the end where it meets the cutoff. When the leak is at the top of the supply tube, all that should be required is the tightening of the nut where the supply tube joins the faucet.

Faucet leaks

Faucet leaks with bidets are basically the same as they are with other common faucets. If the faucet seats are worn, water will drip from the faucet. A faulty washer or cartridge can cause a faucet leak, and bad O-rings and packing material can allow water to leak around the handles of the faucet. The procedures for correcting these problems are covered in Chap. 10.

Spray assemblies

Spray assemblies can leak where their water supply is connected to them. This is usually caused by a washer that has gone bad, but a leak can also occur if the supply nut is not tight. If you are getting a leak in this location, try tightening the nut; if that doesn't do the job, replace the washer.

Transfer valves

Transfer valves have both inlet and outlet connections that can leak. These connections are made in the same way as the connection for a spray assembly. Tighten the leaking nut first, and then replace the washer if necessary.

It is possible for a leak to develop around the stem of the transfer valve. Tightening the packing nut may stop this type of leak quickly. But if it doesn't, remove the steam and inspect the O-rings, packing, and cartridge.

Vacuum breakers

Many plumbers fail to think about the connections at vacuums breakers that may leak. In the case of bidets, the water connection to the vacuum breaker may be made with tubing, a hose, or a direct connection at the fill valve. Any of these types of connections pose potential for leaking.

In the case of tubing and hoses, tightening the connecting nuts will normally stop a leak. It may be necessary to replace a washer or a compression ferrule, but normally a few turns of the nut is all that is needed.

Vacuum breakers that connect directly to fill valves with threaded nipples can have leaks around the threads. This usually results from a lack of thread sealant, nipples that are loose, or sometimes cross-threaded connections. It is also possible that the nipple has deteriorated at the threads.

Hoses and tubing

There are usually several hoses or pieces of small tubing involved in the water connections of bidets. There is also a tee near the fill valve that can present problems. Frequently these connections are made with a nut and a washer, but they are sometimes made with compression nuts and ferrules. In either case, tightening the connecting nuts will normally stop any leaks. If leaks persist after tightening the nuts, you may have to replace a washer or compression ferrule.

As a word of caution, the small tubing used for these connections will crimp easily. When you are working with this tubing, be careful not to bend it too sharply.

Faucet Options

There are a few faucet options when dealing with bidets. One such option is the over-the-rim faucet. This type of faucet setup does not require a vacuum breaker. The faucet simply mounts on the bidet and is angled to provide water to the bowl. This type of faucet uses standard supply tubes.

Another type of faucet for bidets is wall-mounted. With these faucets the faucet handles and the vacuum breaker are mounted at the wall behind the bidet. The transfer valve is mounted on the bidet. The water supply to wall-mounted faucets is made with standard water distribution pipe.

Deck-mounted faucets are popular with bidets. The faucet handles and the transfer valve are all mounted on the deck of the bidet. The vacuum breaker rises above the fixture from behind. Many of these faucets use standard supply tubes for their water supply.

There are a few special variations available in various models of bidets. For example, there are models where the back of the fixture turns up, similar in appearance to a one-piece toilet. With this particular model, the faucet handles are deck-mounted and the transfer valve is mounted on the rise of the back section. The vacuum breaker for this style is concealed behind the back of the bidet. Standard supply tubes are the common method for supplying this type of bidet with water.

Sloppy Pop-Up Connections

Sloppy pop-up connections can be a problem when working with bidets. The piece that connects the lift rod to the pivot rod is pretty long, and it can get out of adjustment. The configuration of the pop-up assembly lends itself to problems.

If the pop-up connections are not adjusted properly, the pop-up may not open or close dependably. I've seen pop-ups on bidets work correctly three times in a row and then fail on the fourth attempt. This type of intermittent trouble can drive a plumber crazy.

When you respond to a complaint for a bidet that either won't hold or won't drain water to a customer's satisfaction, take a close look at the pop-up assembly. And don't try the pop-up just a time or two and accept it. Work the mechanism several times and don't take anything for granted.

Over-the-Rim Leaks

Over-the-rim leaks are possible with bidets that are equipped with over-the-rim faucets. The angle on these faucets is designed to prevent water from spilling over the edge of the fixture, but the problem still occurs at times.

The water from this type of leak is usually on the floor and easy to troubleshoot. There is really not any other type of bidet leak that will cause water to lie on the floor at the sides or front of the bidet.

If the base of the bidet has not been sealed with caulking, over-the-rim leaks can seep under the base of the fixture and appear to be some other type of leak.

I'll admit that over-the-rim leaks are not a common occurrence, but remember, don't rule out any possibility prematurely when troubleshooting.

In General

In general, bidets are just complicated lavatories that bolt to the floor. Many of the same principles used to service and repair lavatories, as described in Chap. 4, apply to bidets.

With bidets behind us, let's move onto the next chapter and study urinals.

9

Troubleshooting Urinals

How much do you know about troubleshooting urinals? Have you ever worked on a trough urinal? How about a stall urinal? Siphon-jet and washout urinals are more common, but many plumbers have never worked on them either. If you are accustomed to doing residential work, you don't have much call for urinal work. Residential plumbers rarely come into contact with urinals, but commercial plumbers work with them frequently.

There are not many complicated parts to deal with when repairing urinals, but this doesn't mean that troubleshooting them is simple. The fixture itself is not complex, but the flush valves for urinals do contain a number of parts that may cause trouble from time to time.

Fortunately, the parts for flush valves are typically available in kits, so entire sections can be rebuilt with relative ease. This reduces the need for identifying the exact part that is defective, and by rebuilding the section completely, the risk of callbacks is reduced.

Types of Urinals

For the benefit of those who are not at all familiar with urinals, let's start this chapter with some descriptions of various types of urinals

Stall urinals

Stall urinals have their bowls recessed beneath the floor level. A typical stall urinal will have about 4 inches of its total height below the floor. They are usually installed in a bed of sand, and they have 2-inch drains. The finished floor slopes toward the bowl of the urinal.

Stall urinals generally have a $\frac{3}{4}$-inch top spud for a flush valve to be mounted in. The water supply for an average flush valve on a urinal has a diameter of $\frac{3}{4}$ inch. Rough-in measurements vary, but the water supply is normally about $11\frac{1}{2}$ inches above the top of the urinal and about $4\frac{3}{4}$ inches to the right of center.

Trough urinals

Trough urinals are mounted on walls and have varying widths. Some are about $3\frac{1}{2}$ feet wide and others stretch to more than $5\frac{1}{2}$ feet wide. A depth of 7 inches is common, and there is usually a little more than a foot between the front lip and the back of the urinal.

The water supply for a trough urinal is typically a $\frac{1}{2}$-inch pipe. The pipe connects to a flush-pipe assembly that runs across the back of the urinal. The flush-pipe sends streams of water down the back of the urinal to clean it.

The trap arm for a trough urinal normally has a diameter of $1\frac{1}{2}$ inches. These urinals use exposed traps, usually P-traps.

Washout urinals

Washout urinals are wall-mounted and drain through a $1\frac{1}{2}$-inch tailpiece; however, some models have 2-inch drains. They have a $\frac{3}{4}$-inch top spud for a flush valve, and often have integral strainers. A standard P-trap is normally used to connect washout urinals to the drainage system.

Blowout urinals

Blowout urinals are wall-mounted, with 2-inch drains. The top spud for this type of urinal has a $1\frac{1}{2}$-inch diameter. Flush valves are required for blowout urinals.

Siphon-jet urinals

Siphon-jet urinals have 2-inch drains and $\frac{3}{4}$-inch top spuds. This type of urinal is wall-mounted.

Similarities

There are many similarities among the many types of urinals. Most urinals are equipped with flush valves. All urinals discharge through traps, and in one way or another, all urinals are set against walls. Integral traps are found on some models, and P-traps are used on others. Some have larger top spuds than others, and some have larger drains than others. While the flushing action of the various types differs, the general concept is the same. Water is introduced at the top of the urinal and leaves at the drain in the bottom of the bowl. Strainers, of one type or another, are found in most urinals.

When it comes to troubleshooting urinals, there is not a lot to look for. Most problems are either related to flush valves or drainage obstructions. Let's start our tour of troubleshooting with the flush valves.

Flush Valves

Flush valves are commonly used to flush urinals. These valves are not very complicated to work on, and many of their repair parts are sold in kits. When the kits are installed, the flush valve is rebuilt and ready for continued service. Let's do a quick rundown on the major parts of a flush valve. We will start where the water supply comes out of the wall and work our way down to the top of the urinal.

The first part of the valve is the supply flange. This part simply covers the inlet pipe for an attractive installation.

The next part we come to is the control stop. This is a key element of the flush valve. It is where the water pressure to the valve is controlled. There is a chrome cover that hides a screwdriver stop in the housing. Once the cover is removed, a screwdriver can be used to open and close the valve. When you need to cut the water off to a urinal, this is where you do it.

The horizontal chrome section between the control stop and the main valve body is called a tailpiece. The tailpiece connects to the body of the valve and carries water from the control stop to the flushing mechanism.

As we move down the body of the valve, we find a handle coupling and a handle. This is the section of the valve that gets the most use.

The long tubular section between the valve body and the urinal is called a flush connection. This is where the vacuum breaker parts are located.

The next part we come to is the spud flange. This trim cover hides the spud coupling that connects the flush valve to the urinal.

There are numerous parts inside flush valves. If you open one up, you will find springs, diaphragms, disks, nuts, washers, and assorted other parts. Luckily, these many parts are available in kits, so you don't have to know that the bad part is the handle spring. All you have to do is rebuild the handle assembly with a repair kit. This way, you are replacing all of the possible defective parts in one service call, and there is little chance of a callback.

The types of repair kits available

Many types of kits are available for rebuilding the working parts of flush valves. Since there are several, let's look at them one at a time.

Inside parts. There is a kit for replacing all the inside parts of the main body's head. These parts include a brass relief valve, a disk, a diaphragm, and a brass guide.

Washer set. The washer set that comes in kit form contains the disk, diaphragm, and washer that are installed between the brass guide and the relief valve.

Push-button assembly. A kit for the push-button assembly is also available. This kit includes numerous parts that will rebuild the entire push-button assembly. Some of the parts included are a handle coupling, socket, and spring. If you don't need to replace all of the push-button parts, you can get one of two kits that will repair portions of the assembly. One kit is available for replacing the push-button arrangement, and another kit will allow you to repair the existing push-button system.

Handle assembly. Kits are available for the complete replacement of handle assemblies. However, if you don't need the whole kit, you can buy individual parts for the handle assembly. There is also a handle-repair kit available.

Vacuum breaker repair. Vacuum breaker repair can be accomplished with a kit of parts.

Spud couplings. Spud couplings are available in kit form, too.

Outlet-coupling assembly. The kits for outlet-coupling assemblies replace existing assemblies and include all needed nuts and washers.

   With the use of these kits, rebuilding a flush valve is fast and easy. Once you locate the area where the problem is likely to be, you can replace the inner workings and be on your way to the next job.

Traps

When you are working with urinals, you may be dealing with integral traps or external traps, which are usually P-traps. There is not much you have to worry about with integral traps, unless they become stopped up. There is, however, a gasket where the fixture outlet meets the drain pipe that could go bad and cause leaking.

External traps can get bumped and slip nuts can come loose. It is also possible for thin-gauge traps to deteriorate. There is not normally much call for problems with the traps of urinals, unless they are stopped up, but there are times when you may have to service the traps, so let's talk about them next.

Integral traps

Integral traps in urinals can be compared to those in toilets. Unless the trap becomes obstructed, there is nothing to be concerned with. The only area where trouble is likely to occur, other than for stoppages, is the gasket that is installed between the urinal outlet and the wall connection.

If you are faced with water leaking from behind a urinal that has a concealed trap, this is the first, and about the only item, you have to check. If the gasket is bad, you can replace it without extreme effort. You will, however, have to remove the urinal from the wall to do the job, and this involves disconnecting the flush valve.

Once the flush valve is loose, you can remove the lag bolts that hold the urinal to the wall and lift it off its bracket. This will give you access to the drain gasket.

External traps

External traps are not rare on urinals. These traps are typically standard P-traps. This creates a possibility for leaks at the trap adapter, the J-bend unions, and the tailpiece connection. While these connections rarely leak after initial installation, they can begin to leak for a number of reasons. For example, slip-nut washers can dry out and allow water to leak past them. Cleaning crews sometimes hit the traps with mop handles and knock connections loose. And, metal traps can deteriorate after exposure to untreated water. Any of these possibilities can create a leak that you may have to fix.

Troubleshooting Questions and Answers

In this section of the chapter, we are going to investigate urinals further with some troubleshooting questions and answers. I'll pose the question, and you try to answer it. You will find the answers following the questions.

1. You are called to a restaurant to repair a urinal. The restaurant owner has explained the job to you in the best way she can. Basically, the urinal is filling up nearly to the flood-level rim before the water goes down the drain.

When you go into the men's room to check out the urinal, you are expecting to find a problem in the trap or the drainage piping. You flush the urinal and the water swells up within the bowl, nearly overflowing. This particular model has an integral trap, so you decide to snake the drain before pulling the fixture off the wall.

You insert a ¼-inch bulb-head spring snake into the drain and work it through the trap. After running 25 feet of snake into the drain, you are sure the problem must be corrected. However, after retrieving the snake and flushing the urinal, the water still swills dangerously close to the flood-level rim. Confused, you put the snake down the drain again. The results are the same.

What would you do next? Do you have any idea of what the problem might be? Well, let's see.

You pull the urinal off the wall for further investigation. When you put an inspection mirror in the trap, you find the problem. A deodorizing disk that had been in the bowl of the urinal has become lodged in the trap, creating a baffle that slows down water being evacuated. Because of the shape and mobility of the deodorizing disk, your snake was able to get past the obstruction.

Once the disk is found, you are able to fish it out of the trap with a piece of wire. You reinstall the urinal and test it several times. The problem is solved.

2. This situation involves a urinal that is not flushing properly. When you depress the handle, water washes the fixture, but with very little force. What is the most likely cause of this problem?

There is a good chance that the screwdriver stop is not open far enough. Removing the stop cover and opening the valve will probably correct the problem.

3. Here, you are working with a trough urinal. The customer has complained that there is very little water pressure rinsing the fixture. The urinal is located in the bathroom at a recreational beach. The water source for the plumbing is a well.

You arrive on the job and can see that the trough is not being washed properly. What would you do first? You would probably check to see that the fixture was receiving adequate water pressure, and this would be a good decision. We will assume that you checked the water pressure at the control valve and found it to be suitable, but the pressure in the fixture is still below par. What should you do next?

Do you think that perhaps the flush tube has become clogged with mineral deposits? This would be my guess. Yes, you remove the flush tube and find that it is partially blocked with mineral deposits. You clean the flush tube and the urinal works correctly.

What might you suggest to the property owner to avoid similar problems in the future? If you thought of suggesting the installation of a water softener, you are correct.

4. This question deals with a urinal that is flushing improperly. The flush valve is not consistent in its flushing action. Sometimes it flushes pretty well, and other times it hardly flushes at all.

When you arrive on the job and flush the urinal, it works okay except that you notice the handle seems sloppy. After several flushes you experience the intermittent dependability of the flush valve. What do you think will correct this problem?

Installing a kit to rebuild the handle assembly is your best bet. By rebuilding the assembly, you can feel secure in your chances of correcting the problem.

5. This urinal is equipped with a standard urinal self-closing valve. The fixture is about 5 years old. The complaint is that water constantly dribbles into the bowl, resulting in high water bills. What can you do to fix it? This is a situation where the self-closing mechanism has become worn and must be rebuilt or replaced.

6. Here, you are called to an old school, where water is leaking onto the ceiling of the first floor. The maintenance director shows you the area where water is leaking through the ceiling and explains that the leak is not constant, but that it does occur often.

Since the leak is not constant, you rule out the water distribution system. With your attention focused on the drainage system, you move upstairs to check out the bathrooms. One of the bathrooms contains water closets, lavatories, and urinals. The other restroom contains only water closets and lavatories.

A quick inspection reveals no obvious signs of where the leak might be coming from. You flush the toilets and look for water that may seep out from around their bases. No water appears. The urinals are stall urinals. You look at their bowls to see if any cracks are present. None are. After having flushed all the toilets, you return to the first floor and look for dripping water. After a few minutes of watching and not seeing any water, you go back up to the bathrooms.

This time you flush the urinals and go back downstairs. When you enter the room where the ceiling has been damaged you see water dripping through the ceiling. You seem to have narrowed the leak down to being one of the urinals. Now the question is which urinal is it, and what is making it leak.

The ceiling is already badly damaged, so you cut a small hole in it to drain off excess water. When the dripping has stopped, you return to the bathroom and flush the first urinal. Then you go back downstairs and wait for water, but none comes. After a few minutes, you repeat the process with the second urinal. This time water does leak out of the ceiling. Now you know that a urinal is definitely the cause of the leak, and you have identified which urinal is responsible for the problem.

You return to the urinal that is causing the trouble and give the bowl a close inspection. Everything appears normal. Where do you think the leak is coming from? Did you guess the drain? If you did, you're right. Now the question is, how do you fix it?

Since the stall urinal is recessed into the floor by several inches, you don't have easy access to the drain connections. This job will require you to cut the ceiling out from below the urinal. Once you gain access to the drain, you see the leak is coming from the drain connector. You repair the leak and test the urinal. Everything remains dry, and you have solved the problem.

The Process of Elimination

The process of elimination is very effective when troubleshooting urinals. Since there are relatively few things that can go wrong with a urinal, it doesn't take long to get to the root of a problem.

The drainage system of urinals is simple, and there are only a few possibilities for problems. The drain can become blocked. A trap can cause a leak, and with some models the gasket between the urinal and the wall receptacle can be a problem.

Stoppages are easy to identify and are usually not difficult to correct. Leaks at traps are easy to see and normally simple to fix. Gasket leaks can be a bit more difficult to find and fix, but even they don't rank high on the list of hard-to-fix plumbing problems.

Urinals equipped with simple, self-closing valves don't offer many problems to plumbers. If there is trouble with the valve, it is easy to replace.

Flush valves contain more parts and present more potential for trouble than standard self-closing valves, but they are still relatively simple to troubleshoot. With repair kits available for the various sections of flush valves, rebuilding a faulty valve is not a big job.

Now that we are done with urinals, let's move on to the next chapter and learn about the troubleshooting procedures for faucets and valves.

10

Troubleshooting Faucets and Valves

When you want to know about troubleshooting faucets and valves, you must be willing to take in a lot of information. There are so many different styles, types, brands, and models of faucets and valves on the market that it takes a lot of ambition to be able to troubleshoot all of the various situations you may run into.

Since there are so many possibilities for problems with faucets and valves, we are going to have to spend a substantial amount of time on this issue. I'm going to give you detailed explanations, case histories, and numerous drawings to aid you in developing your troubleshooting capabilities.

What types of techniques are we going to study? We will look at all the requirements involved in finding and fixing problems with faucets and valves. We will talk about lavatory faucets, laundry faucets, yard hydrants, tub and shower valves, and much more. This comprehensive chapter is going to prepare you for all the faucet and valve problems you are likely to encounter in the field.

Kitchen Faucets

Kitchen faucets come in many shapes and sizes. There are wall-mounted faucets, single-handle faucets, two-handle faucets, faucets with spray attachments, faucets without spray attachments, and specialty faucets. These faucets can have 6- or 8-inch centers. They can have one-piece bodies, or they can have individual components that must be put together to make a working faucet. Clearly, there are many possibilities for all sorts of problems with kitchen faucets.

Faucet washers

There was a time when faucet washers were found in all faucets. Times have changed and so have the inner workings of faucets. Today, faucets don't always have washers or bibb screws in them. Many modern faucets have springs, rubber seals, cartridges, and washerless stems. Gone are the days of putting an assortment of washers, screws, and seats on your truck to take care of any problem that might come along with a dripping faucet.

While the need for many more parts has evolved over time, the principles behind what makes a faucet leak are still pretty much what they were 10 or 20 years ago. We will get into these principles and practices in just a little while.

Washerless faucets

Washerless faucets are kind of like computers; they are great when they work the way they are supposed do, and they are horrendous when they don't. With washerless faucets, it is necessary to replace the entire stem unit. These stems can cost nearly as much as a replacement faucet would cost.

Cartridge-style faucets

Cartridge-style faucets are my favorite. I like these faucets because they are so simple to rebuild. It also happens that I've used one particular brand of cartridge-style faucet for about 15 years with almost no warranty callbacks, and this means a lot to me or any other plumbing contractor. There is also a side benefit to my favorite cartridge-style faucet; if for any reason the hot and cold water is piped to the wrong side of the faucet, the cartridge can be rotated to correct the problem. This is much easier than crossing supply tubes or pipes, and it doesn't show as a mistake on the plumber's part.

Ball-type faucets

Ball-type faucets are very common and popular. This style of faucet is not difficult for plumbers to repair, but there are many individual parts to be concerned with. Unlike cartridge-style faucets, where there is only the cartridge to replace, ball-type faucets have the ball, springs, and rubber parts that can all give you trouble.

Disposable faucets

Disposable faucets are becoming more and more a part of today's plumbing. Plumbers used to fix faucets, but now they often just replace them. Some old-school plumbers say this is done because the new crop of plumbers don't have the knowledge or desire to repair faucets. I disagree with this generality. Being somewhat of an old-school plumber myself, I can understand what the other veterans are talking about, but I can also see the younger plumbers' points of view.

Many modern faucets are not worth repairing. In fact, there are many faucets available that cost less than a replacement stem for more expensive faucets. For example, I can buy a complete lavatory faucet, including pop-up assembly, for less than $10. When you consider that stems can run anywhere from $6 to $17 a piece, why would you repair an old faucet when it would be cost effective to replace the entire unit? The customer gets a new faucet, and the plumber reduces the risk of a callback from a repair that doesn't last.

Troubleshooting Kitchen Faucets

Now that we have touched on the basic types of faucets in use today, let's devote some time to troubleshooting and repairing the various types. Much of what you are about to learn can be applied to more than just kitchen faucets. Many of the same troubleshooting techniques can be used for any type of faucet, and many of the repair options will be similar to those available for other types of faucets. Whether you are trying to eliminate a drip in a kitchen faucet, a lavatory faucet, a bar sink faucet, or a deck-mounted laundry faucet, most of the work will be closely related. Now with that in mind, let's jump right into the troubleshooting and repair of kitchen faucets.

Dripping from the spout

What's wrong when there is water dripping from the spout of a kitchen faucet? Usually, the problem is either with the faucet seat or the faucet stem or washer. However, the problem is not always caused by these parts being defective. Are you wondering how the faucet could be dripping if the parts are not defective? You should be, and that's good; it means you are thinking. I will explain how the problem could be caused by something other than defective faucet parts in a few moments.

When you walk in to troubleshoot a dripping faucet, you must first determine what type of faucet you are dealing with. In other words, does the faucet have washers, a ball assembly, a cartridge, or some other type of mechanism? Experienced plumbers will know the answer to this question as soon as they see the faucet in many cases. If you can't tell by looking at the exterior of the faucet, it is easy enough to determine the type by removing the handle assembly, and you're going to have to do this anyway, so even rookies are not at a great disadvantage. A visual inspection of the faucet seats, washers, and/or stem will tell you what the problem is, most of the time.

An exception.   There is an exception, the one I mentioned just a few moments ago. Sometimes grit gets between the seat and the stem washer. When this happens, the stem washer cannot seat properly, and a drip results.

Plumbing systems in which iron particles are present are the ones most likely to experience a drip without having defective plumbing parts. Gravel, sand, and other impurities can also keep a faucet from closing fully. The debris in the faucet will usually be visible, but sometimes it will be small enough to avoid detection. If you suspect the leak is being caused by debris, flush the stem hole out with water. You can do this by turning on the water to the faucet slowly while the stem is removed. Water will bubble out of the hole and flush the grit out. Don't turn the water on too fast, or you will create a geyser that can shoot all the way up to a ceiling.

Faucet seats.   Faucet seats sometimes become worn or pitted. When this happens, faucets drip. Small grinding wheels can be used to resurface the seats, but it is usually better to replace the damaged parts.

Corrosive elements in the water, such as iron and acid, are common causes of deteriorating faucet seats. Finding such problems in a plumbing system opens the door for the potential sale of water conditioning equipment.

Washers.   Washers in faucets can wear out or be damaged by rough faucet seats. If a seat has become pitted and rough, it can cut a washer to the point where the faucet will leak. If you inspect faucet washers and find them punctured or cut, you should replace the faucet seats at the same time you replace the washers.

If the bibb screws that hold the washers to the stems are weak, you have found evidence of corrosive water conditions. This gives you yet another opportunity to recommend water treatment equipment to the customer.

Washerless stems.   Washerless stems are not considered repairable. When these components go bad, they should be replaced. It may be possible to extend their life a short time by sanding the bottom of the stem with a light-grit sandpaper, but you cannot expect long-term success with this method.

Single-handle faucets.   Single-handle faucets that are dripping are going to require some replacement parts unless the drip is being caused by debris lodged in the faucet. The parts could be a single cartridge, O-rings, springs, ball assemblies, or rubber seals. It is usually worthwhile to rebuild these types of faucets completely to avoid the risk of unwanted warranty callbacks.

Leaking around the base of a spout

When water is leaking around the base of a spout from a kitchen faucet, you can bet on the problem being a bad O-ring. You don't need much troubleshooting ability to solve this problem. All you have to do is remove the spout and replace the O-ring or O-rings, whichever the case may be.

Leaking around a faucet handle

When water is leaking around a faucet handle, you have one of two problems. Either the packing in the stem has gone bad, or there is a gasket or O-ring around the stem that is defective. Once you remove the handle from the faucet, you will be able to determine which problem you are faced with.

If water is leaking out around the base of the stem, you have a gasket or O-ring problem. When the water is coming out around the part of the stem that turns when the faucet is used, the problem is with the packing. Neither of these problems is difficult to correct.

With the water cut off, you can remove the stem and replace the gasket or O-ring quickly. If the problem is with the packing, you will have to remove the stem assembly and install new packing material. In either case, the job will take only a few minutes to complete.

Won't cut off

What should you look for when the water from the faucet won't cut off? Well, the first thing to do is to cut off the water supply to the faucet. With that done, you can proceed with your troubleshooting. What are you likely to find? You may find a large piece of grit between the faucet seat and the stem washer. It is also possible that the bibb screw has deteriorated and allowed the washer to float about, resulting in a steady stream of water that won't cut off. A quick look into the faucet body should tell you what is causing the problem.

Poor water pressure

Poor water pressure is a common complaint. When the pressure is low at only one fixture, the troubleshooting skills needed to solve the problem are not extensive. Before you begin a massive investigation into the inner workings of the troublesome faucet, there are a few simple things to check.

Start your troubleshooting by removing the aerator from the faucet. Look at the screen and diverter disk to see if they are blocked by debris or mineral deposits. Turn the faucet on and see what the water pressure is like with the aerator removed. Many times this will prove to be the simple solution plumbers hope for. If the aerator is the culprit, all you have to do is clean or replace it.

When the aerator is not at fault, check the cutoff valves under the fixture. Make sure the valves are in the full-open position. You might be surprised at how often the valves are partially closed, causing low water pressure at faucets.

While you are checking the cutoffs, look at the supply tubes running up to the faucet. If a supply tube has a bad crimp in it, you have probably found the cause of the low water pressure. Also, see what type of material has been used for the water distribution piping.

If the water distribution piping is made of galvanized steel, you probably have your work cut out for you. Old galvanized water piping tends to clog itself up with rust and mineral deposits. As the pipe slowly clogs, the water pressure to the plumbing fixtures is gradually decreased. The reduction is not noticeable in the early stages, but in advanced stages the flow can be reduced to a trickle. This type of problem requires extensive work. The old pipe must be removed and new piping installed.

Once you have eliminated the external causes of low pressure, you must look inside the faucet. You may find debris blocking the faucet inlets or delivery tubes. One way to be sure the problem is in the faucet is to disconnect a supply tube and turn the water on. If you have good pressure at the tip of the supply tube, you know the problem has to be in the body of the faucet.

It may be necessary to replace some faucets that have become clogged with mineral deposits or debris. In other cases you can simply clear the blockage and put the faucet back into service.

Not a uniform flow

When a faucet is delivering water that is not in a uniform flow, the problem is with the aerator. A spraying or rough stream of water is a sure indication of a partially plugged aerator. Remove the aerator and clean or replace it.

No water at the spray head

Many plumbing customers complain that they have no water at the spray head of their kitchen faucet. There are only a few possible causes for this type of problem.

When you have a spray attachment that will not deliver water, check the spray hose. These hoses frequently become kinked under sinks. A bad kink can render the spray head useless.

If the spray hose is in good condition, take the spray head apart and inspect for mineral deposits or debris. You will often find iron deposits blocking up the works. If necessary, replace the spray head.

I can't recall ever having it happen, but it is possible that the outlet port on the bottom of the faucet could become blocked, cutting off the water supply to the spray hose. If you have replaced the old spray assembly with a new one and are still not getting water, consider the possibility of a blockage at the connection port.

Limited water at the spray head

Limited water at the spray head is more common than a complete lack of water. The same troubleshooting steps offered for solving no-water problems can be followed to solve problems with limited water at the spray head.

Water dripping from beneath a faucet

Water dripping from beneath a faucet can be frustrating. This type of problem is often misdiagnosed as being a leak at the supply tube. If water leaks from the bottom of a faucet and runs through the faucet holes in the sink, it can appear to be coming from the supply-tube connection.

Acidic water is a primary cause of leaks in the bodies of faucets. The thin copper tubing used to mix and deliver water in the bodies of faucets can be eaten up by water with a high acid content. Small pin-holes can develop and allow water to escape within the faucet housing. You won't be able to put your finger on these leaks without first removing the entire faucet from the sink.

When you have a pitted or corroded faucet body, you should replace it. While it is sometimes possible to repair the leak with a spot-soldering job, you will be better off to replace the faucet.

Troubleshooting Other Faucets

Lavatory faucets

Lavatory faucets are very similar to kitchen faucets when it comes to troubleshooting and repair work. There are, however, more multipiece lavatory faucets used than there are multipiece kitchen faucets. Multipiece faucets are faucets that consist of individual handles and spouts that are connected, by plumbers, with tubing beneath the fixture. While single-body faucets are much more common, multipiece faucets are frequently found in more expensive plumbing systems. These faucets offer a few wrinkles that single-body faucets don't.

Most of the troubleshooting techniques used for multipiece faucets are the same as those used for single-body faucets. However, the connection points under fixtures for multipiece faucets can produce leaks that don't exist in single-body faucets.

Most multipiece faucets (Fig. 10.1) are connected with small copper tubing and compression fittings. The tubing is soldered into the individual faucet components and connected to the mixing spout with compression fittings. The tubing is so small and thin that it crimps easily. This can result in low water pressure. It is also easy to break the soldered connections between the tubing and the individual components.

When you are working with multipiece faucets, be careful. Too much stress on the components can twist, crimp, or break the tubing connections. Since multipiece faucets are typically expensive, you don't want to have to replace one at your own expense.

In the interest of space and time, I will not give a blow-by-blow

Figure 10.1   Three-piece, widespread lavatory faucet. (*Courtesy of Moen, Inc.*)

Figure 10.2   Single-handle lavatory faucet. (*Courtesy of Moen, Inc.*)

account of how to work with lavatory faucets. You can use the same principles and techniques described for kitchen faucets to work with lavatory faucets (Figs. 10.2 through 10.5).

Bar sinks and laundry tubs

Deck-mounted faucets for bar sinks and laundry tubs fall into the same troubleshooting and repair category as kitchen faucets. As with lavatory faucets, refer to the section on kitchen faucets for information on troubleshooting the faucets for bar sinks and laundry tubs.

Troubleshooting Bathtub and Shower Faucets

The faucet valves for bathtubs and showers are a little different from the faucets used for kitchen sinks (Fig. 10.6). The tools and techniques required for troubleshooting these valves are similar, but not always the same. Let's see how the differences affect your work.

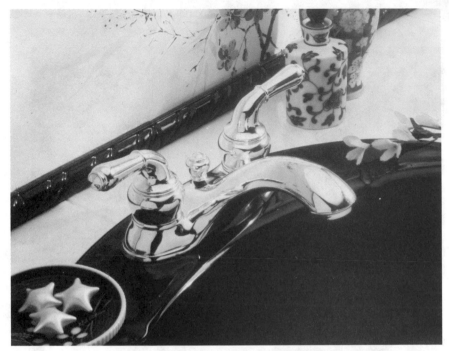

Figure 10.3   Lever-handle lavatory faucet. (*Courtesy of Moen, Inc.*)

Figure 10.4   Mini-widespread lavatory faucet. (*Courtesy of Moen, Inc.*)

Figure 10.5   Lever-handle lavatory faucet. (*Courtesy of Moen, Inc.*)

Dripping from the spout

When you have water dripping from a tub spout or a shower head, you can use the same basic principles described for kitchen faucets in your troubleshooting and repair work. You may, however, need a set of tub wrenches to remove the valve stems. Tub wrenches (Fig. 10.7), for those of you who aren't familiar with them, resemble deep-set sockets.

Two- and three-handle tub and shower valves are often equipped with valves that have their wrench flats concealed in the finished wall. Since you can't get a standard wrench on the flats to remove the stems, you must have tub wrenches to reach into the wall and remove the parts.

Once you have the stems for tub and shower valves removed, the rest of the troubleshooting and repair procedure is comparable to that used for kitchen faucets.

Leaking around a faucet handle

When water is leaking around a faucet handle on a tub or shower valve, the cause of the leak is usually defective packing around the

Figure 10.6 Pressure balancing tub/shower valve.
(*Courtesy of Symmons Industries, Inc.*)

Figure 10.7 Tub wrenches.

stem. The stems for tub and shower valves are larger than those found in sink faucets, but they share many of the same characteristics. By removing the stem, you can replace the bad packing. It is possible that the leak is coming from a bad O-ring, so check this before reinstalling the stem.

Won't cut off

If you have a tub or shower valve that won't cut off, you can apply the same basic principles and procedures used for kitchen faucets in your troubleshooting and repair work.

Poor water pressure

Poor water pressure is rare with bathtub valves, but shower heads sometimes suffer from a lack of desirable water pressure. The cause is usually a mineral buildup in the water-saver disk that is housed in the shower head. If you remove the head and check the disk for obstructions, you will probably find some. A quick rinse under a faucet will often clear the debris. In stubborn cases you may have to poke the pieces out with a thin piece of wire or replace the disk. It is also conceivable that the holes in the shower head itself are plugged.

Uniform flow

A uniform flow is another problem you could encounter with a shower head that is partially plugged with foreign objects. While tub spouts won't give you this type of trouble, shower heads can. When you have an erratic spray from a shower head, check the head and water-saver disk for blockages.

Diverters that don't divert well

Diverters that don't divert well are a fairly common problem with three-handle tub-and-shower valves. In some cases water isn't divert- ed from the tub spout to the shower head. At other times water will run out of both the tub spout and the shower head simultaneously. The cause of these types of problems is the stem washer or the valve seat. Treat the condition just as you would a dripping faucet, and replace the seat or washer as needed.

Wall-mount faucets

Wall-mount faucets are not very different from their deck-mounted cousins when it comes to troubleshooting and repairing them. You can apply the same principles and procedures discussed for kitchen faucets to your work with wall-mount faucets.

Frost-free hose bibbs

Frost-free hose bibbs are similar to a sink faucet in terms of trou- bleshooting. The stem for a frost-free hose bibb is, however, much longer than a stem for a standard faucet. The fixture still contains a seat and a washer, but the stem is extra long. The length of the stem depends on the length of the hose bibb, but it can easily be up to a foot long.

Hose bibbs also have packing and packing nuts that can spawn leaks. You can apply the basic troubleshooting principles and prac- tices we discussed in the section on kitchen faucets to frost-free hose bibbs.

Learn the Basics

Once you learn the basics of troubleshooting valves and faucets, you will be able to tackle just about any type of unit with a good deal of success. The general principles are about the same for all types of faucets and valves. The job only gets tricky when you fail to follow a systematic troubleshooting approach.

As routine as troubleshooting faucets is, the task can become frustrating. There will no doubt be times when the problem doesn't know it is supposed to go by the book. When this happens, you must depend on the experience and knowledge you have gained through field work and study to solve the problem. To drive this point home, let's take a little time to look at some scenarios that you might run into out in the field.

Field Conditions

Field conditions can play a major role in the level of difficulty a plumber faces in solving on-the-job problems. Repair work that should be simple can become quite complicated when there are space limitations. For example, a multipiece faucet that should be simple to repair can become a nightmare if the faucet is offset to one corner of a sink and drawers in the cabinet block your access. The same could be said for trying to replace a washer when the bibb screw has been eaten up by acidic water.

The act of replacing a faucet washer is certainly not complicated. Under normal conditions, almost any adult should be capable of handling the work. If, however, the head of the bibb screw falls apart when you touch it with a screwdriver, the simple job turns into a challenging assignment. If this were to happen to you, what would you do?

When the head of a bibb screw disintegrates, you have a few options. The least complicated option is the replacement of the faucet stem. Some customers, however, will not want to pay for a whole new stem. If customers knew how much they were going to spend for the time it takes a plumber to work with a broken screw, they would be more inclined to replace the stem, and this is something you may want to mention to cost-conscious customers.

If the customer insists on having you spend time working with the broken screw, accept the challenge and overcome it. There are two fairly simple ways of doing this. You can take a knife and cut out the old washer. This will expose the threaded portion of the old screw. Many times the shaft of the screw will not have rotted to a point where it will not turn. If you put a pair of pliers on the threaded shaft, once the washer is removed, there is a good chance you can turn the shaft out of the stem.

Should the shaft of the screw break off, don't panic. Use a small drill bit to drill out the old screw. Then you can either try to rethread the stem, or you can install snap-in washers. Snap-in washers are fitted with two ears that can be pushed into the hole you drilled. Once the ears are in the hole, their natural tension will hold the washer in place.

Every experienced plumber has a personal way to deal with a broken bibb screw. What is the right way to get the job done? Any approach that works is acceptable. You should strive for a procedure that is fast, cost-effective, dependable, and durable and that will satisfy the customer; when you meet this criteria, you have done a fine job.

Field conditions can put some plumbers into tailspins. If they have not spent a considerable amount of time doing real plumbing, they can be mystified by the actions of other plumbers, homeowners, and even inanimate plumbing. This type of situation can arise around any type of plumbing, not just faucets and valves.

I've been on countless jobs where the field conditions made doing my job all but impossible. Even now, with 20 years of experience, I still encounter circumstances in the field that test my abilities to the limits. For example, what kind of a person would install a water heater with the element access doors facing a solid wall? Obviously, someone who had no intention of ever working on the water heater again. I saw an installation like this just last week.

Why would someone thread a coupling onto the supply-tube inlets of a faucet? One of my recent jobs had such an arrangement, and there was a leak between the faucet inlet and the supply tube. If the coupling had not been installed, it would not have been possible for such a leak to exist. The coupling was not needed, and it was not standard plumbing procedure to install it. To make matters worse, the coupling had been installed so tightly that it took a basin wrench with a 14-inch pipe wrench attached to its handle to remove the couplings.

What would inexperienced plumbers think when seeing this strange coupling in a place where the books never show such a fitting? They would probably be confused, and they might try to salvage the fitting or replace it. With my experience I knew the coupling was useless, so I removed it and installed a slightly longer supply tube. The leak was fixed.

The two most important traits to develop when working in the field are experience and product knowledge. Experience only comes with time, but product knowledge can be learned as quickly as you are willing to put forth the effort. If you know how plumbing fixtures, faucets, and devices are put together, you are well on your way to being an excellent troubleshooter. On the other hand, if you do not know that a particular faucet should have a gasket in a particular place, how will you know what the problem is if the gasket is missing?

All major plumbing manufacturers are very willing to provide professionals with detailed information about their products. If you take the time to obtain cut sheets from manufacturers, you can begin your study of what makes faucets and valves work. This same approach applies to all plumbing parts, devices, and fixtures. All you have to do is ask your supplier for the information you want. If the supplier doesn't have detailed drawings, you can get the manufacturer's address from the supplier and request the information directly from the company that makes the plumbing parts. Let me give you some examples of the types of drawings I'm talking about.

Some Examples of Faucets and Valves

Here are some examples of popular faucets and valves for you to look over. As you will see, the drawings provide a lot of detail and information that can make troubleshooting and repair work much simpler. Study these illustrations and become familiar with the inner workings of the faucets and valves; this is one of the most effective ways to begin building the foundation for your troubleshooting skills. (Figs. 10.8 through 10.55)

Now that you have had an opportunity to look over the cut sheets on faucets and valves, let's move on to the next chapter and investigate plumbing appliances.

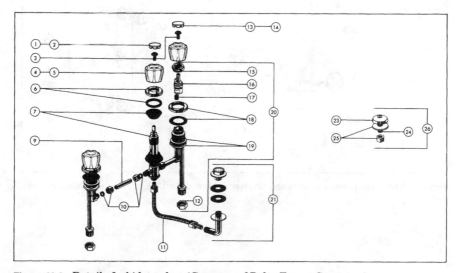

Figure 10.8 Detail of a bidet valve. (*Courtesy of Delta Faucet Company.*)

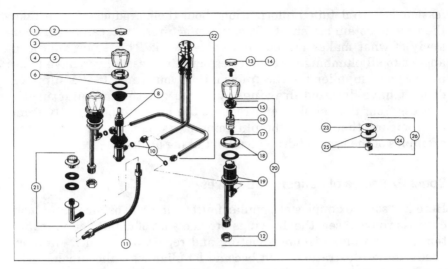

Figure 10.9   Detail of a bidet valve. (*Courtesy of Delta Faucet Company.*)

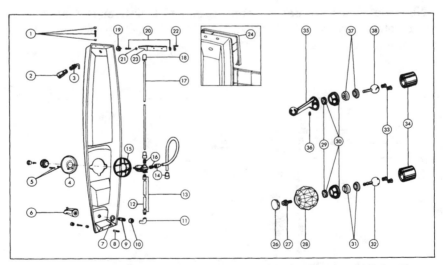

Figure 10.10   Nonpressure balance shower valve detail. (*Courtesy of Delta Faucet Company.*)

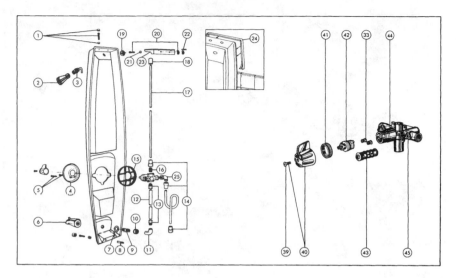

Figure 10.11   Pressure balance shower valve detail. (*Courtesy of Delta Faucet Company.*)

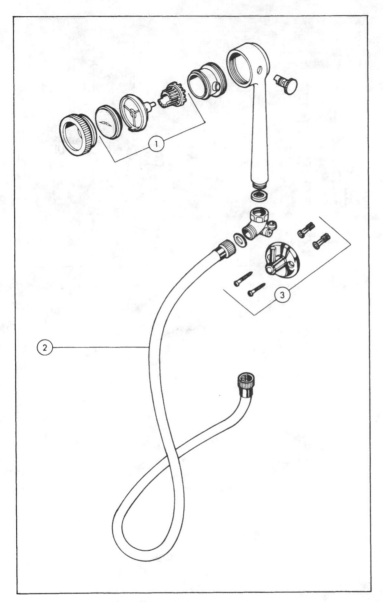

Figure 10.12   Handheld shower. (*Courtesy of Delta Faucet Company.*)

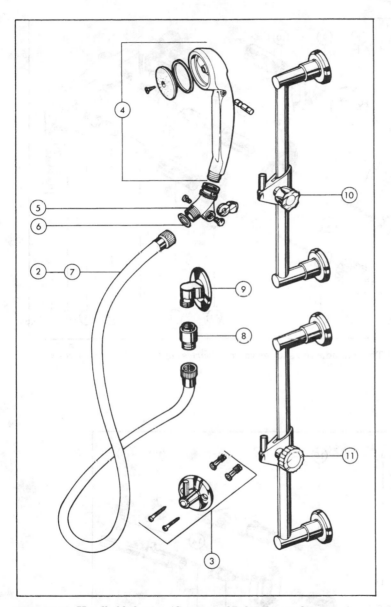

Figure 10.13   Handheld shower. (*Courtesy of Delta Faucet Company.*)

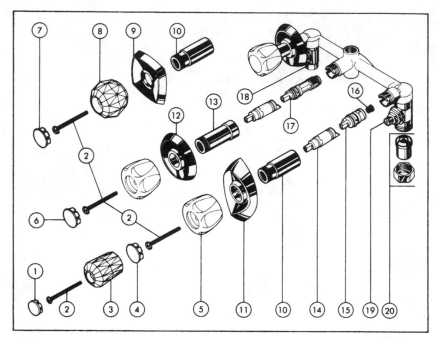

Figure 10.14   Three-handle tub/shower valve. (*Courtesy of Delta Faucet Company.*)

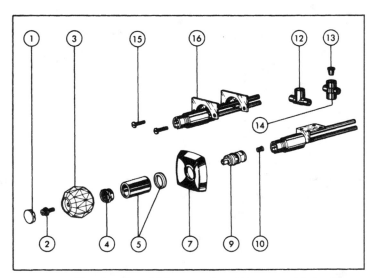

Figure 10.15   Two-handle tub/shower valve. (*Courtesy of Delta Faucet Company.*)

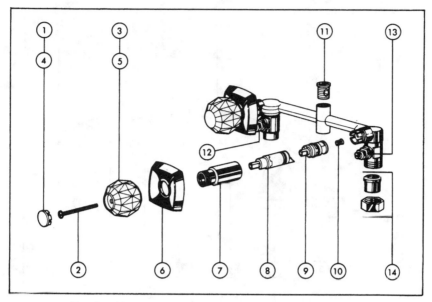

Figure 10.16   Two-handle tub/shower valve. (*Courtesy of Delta Faucet Company.*)

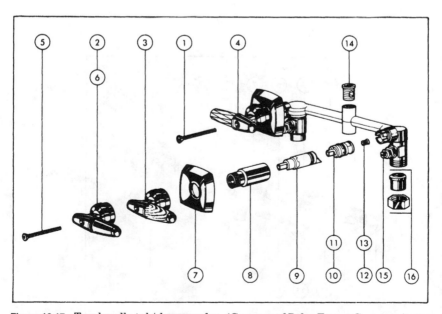

Figure 10.17   Two-handle tub/shower valve. (*Courtesy of Delta Faucet Company.*)

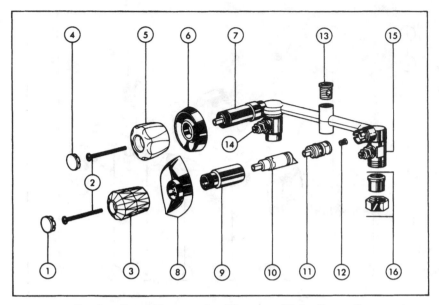

Figure 10.18 Two-handle tub/shower valve. (*Courtesy of Delta Faucet Company.*)

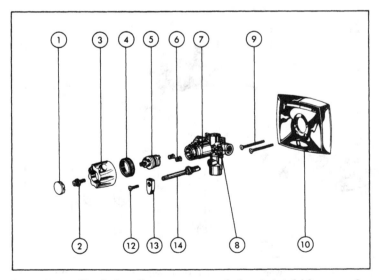

Figure 10.19 Single-handle, nonpressure balancing tub/shower valve. (*Courtesy of Delta Faucet Company.*)

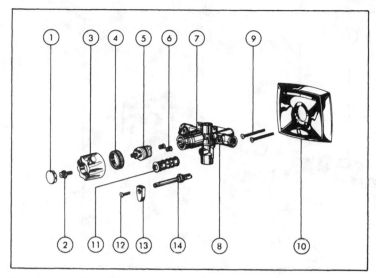

Figure 10.20   Single-handle, pressure balancing tub/shower valve. (*Courtesy of Delta Faucet Company.*)

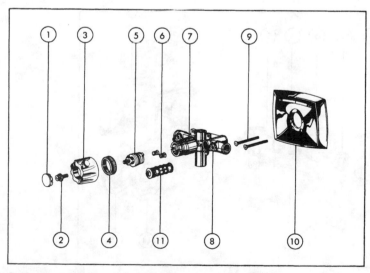

Figure 10.21   Single-handle, pressure balancing tub/shower valve. (*Courtesy of Delta Faucet Company.*)

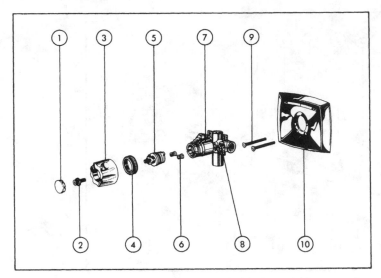

Figure 10.22   Single-handle, nonpressure balancing tub/shower valve. (*Courtesy of Delta Faucet Company.*)

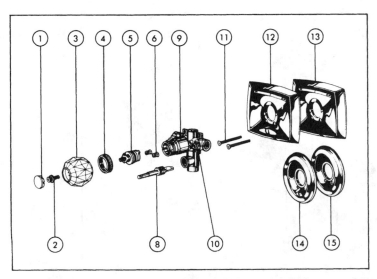

Figure 10.23   Single-handle, nonpressure balancing tub/shower valve. (*Courtesy of Delta Faucet Company.*)

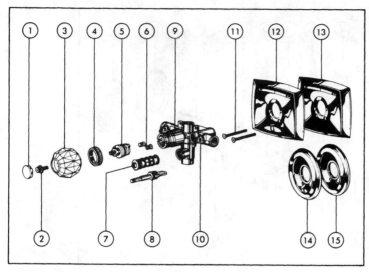

Figure 10.24 Single-handle, pressure balancing tub/shower valve. (*Courtesy of Delta Faucet Company.*)

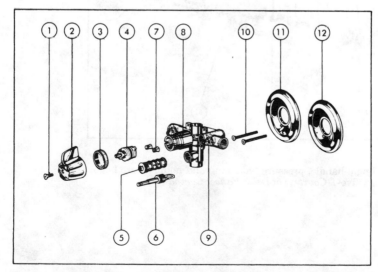

Figure 10.25 Single-handle, pressure balancing tub/shower valves. (*Courtesy of Delta Faucet Company.*)

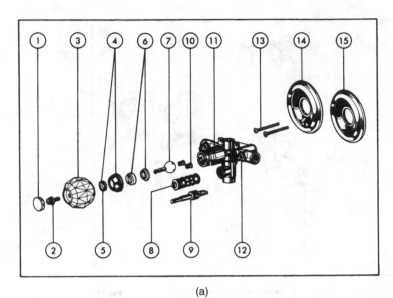

(a)

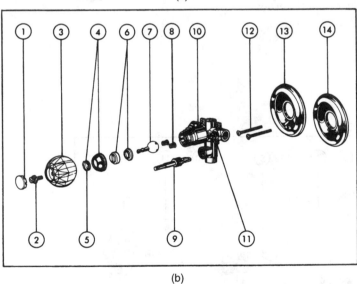

(b)

Figure 10.26   Single-handle, pressure balancing and nonpressure balancing tub/shower valves. (*Courtesy of Delta Faucet Company.*)

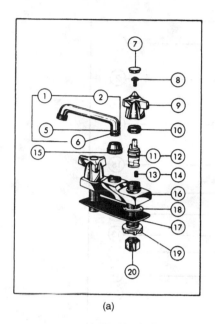

(a)

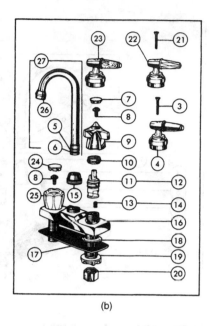

(b)

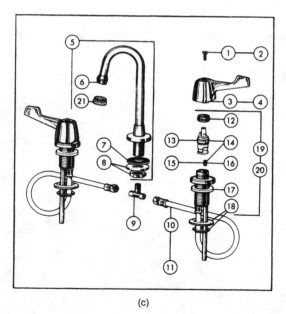

(c)

Figure 10.27   (a) Laundry faucet; (b) bar sink faucet; (c) widespread utility faucet. (*Courtesy of Delta Faucet Company.*)

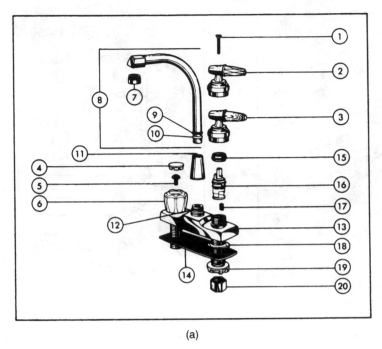

(a)

Figure 10.28a Waterfall lavatory faucet. (*Courtesy of Delta Faucet Company.*)

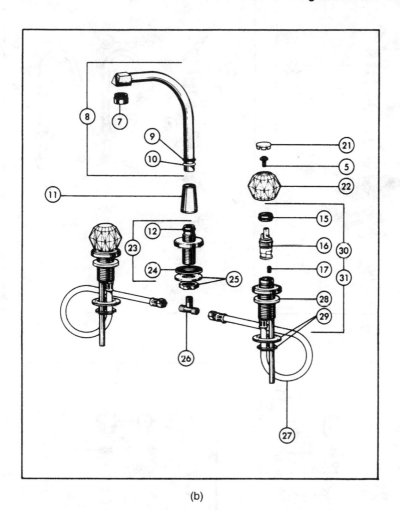

(b)

Figure 10.28b Waterfall, widespread lavatory faucet. (*Courtesy of Delta Faucet Company.*)

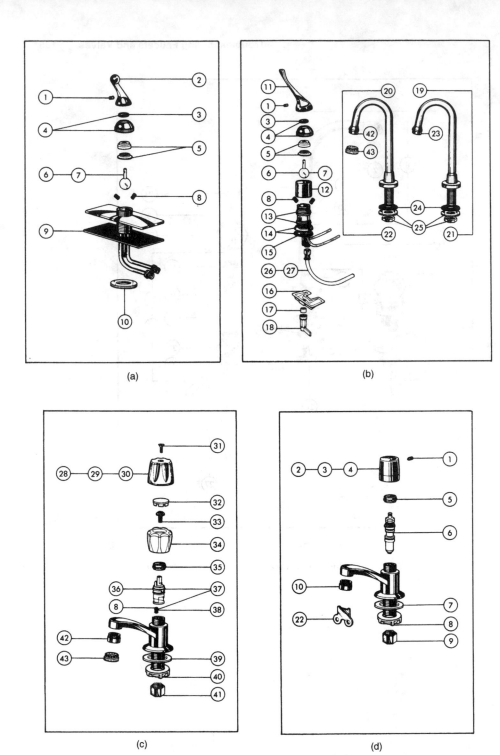

Figure 10.29 (a) Single-handle shampoo faucet; (b) single-handle utility faucet; (c) basin faucet; and (d) slow self-closing basin faucet. (*Courtesy of Delta Faucet Company.*)

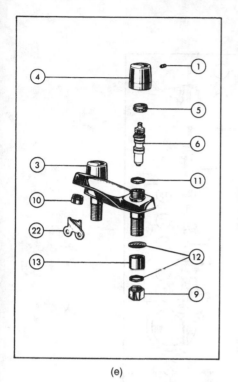

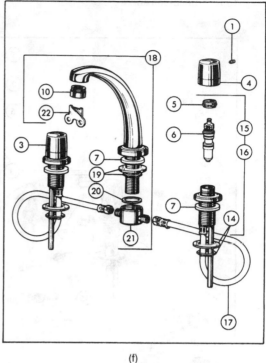

(e)

(f)

Figure 10.29 (*e*) Slow self-closing lavatory faucet; (*f*) slow self-closing, widespread lavatory faucet. (*Courtesy of Delta Faucet Company.*)

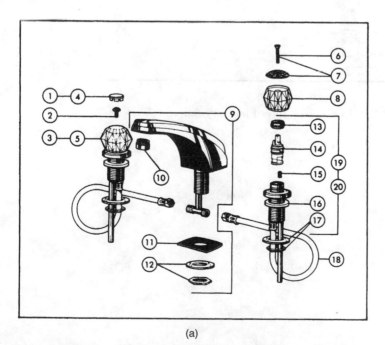

(a)

Figure 10.30*a* Two-handle widespread lavatory faucet. (*Courtesy of Delta Faucet Company.*)

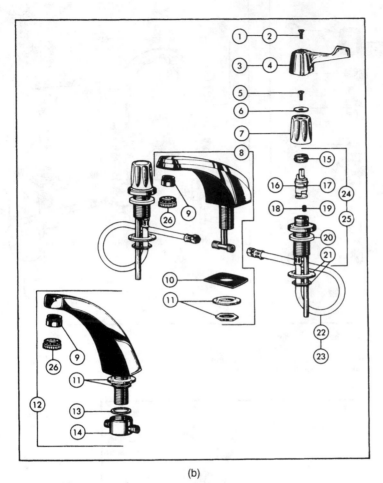

(b)

Figure 10.30b Two-handle widespread lavatory faucet. (*Courtesy of Delta Faucet Company.*)

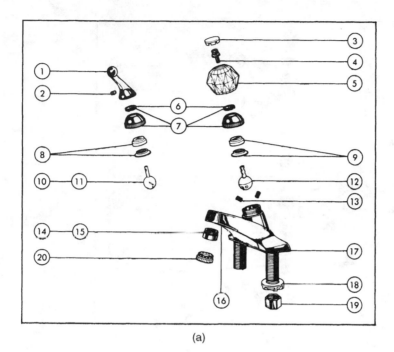

(a)

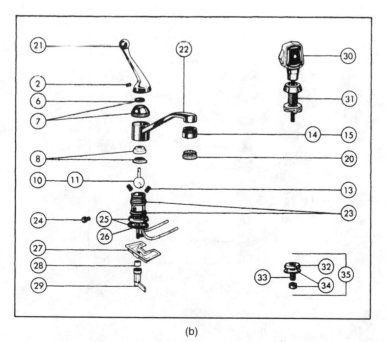

(b)

Figure 10.31 Single-handle, single-mount faucet. (*Courtesy of Delta Faucet Company.*)

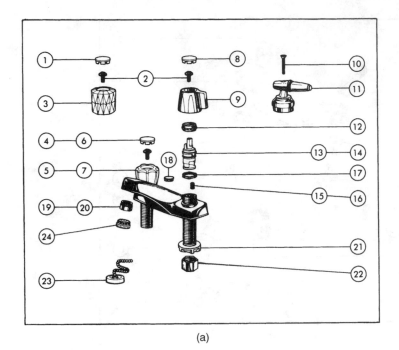

(a)

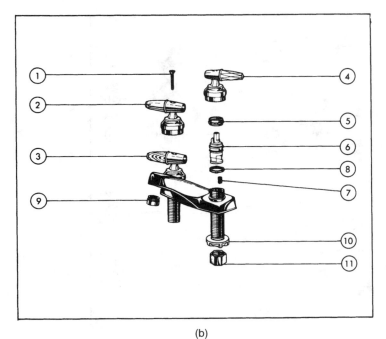

(b)

Figure 10.32 Two-handle lavatory faucet. (*Courtesy of Delta Faucet Company.*)

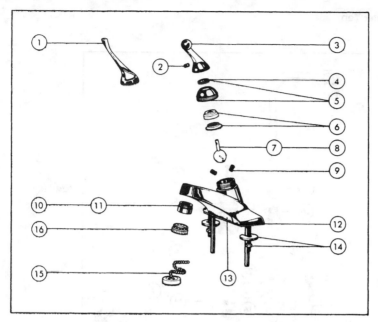

Figure 10.33 Single-handle lavatory faucet. (*Courtesy of Delta Faucet Company.*)

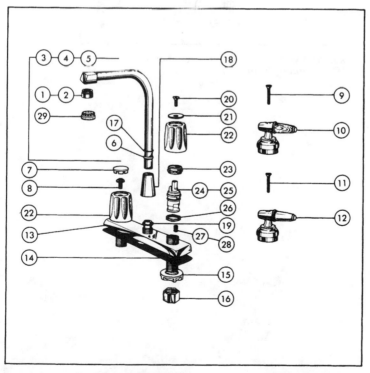

Figure 10.34 Two-handle kitchen deck-mounted waterfall faucet. (*Courtesy of Delta Faucet Company.*)

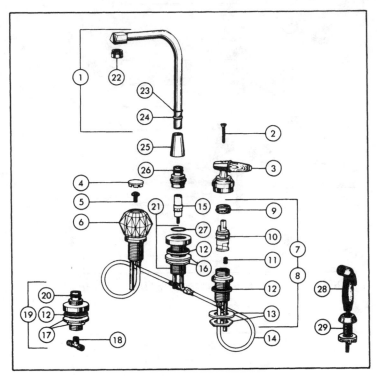

Figure 10.35 Widespread two-handle kitchen deck-mounted waterfall faucet. (*Courtesy of Delta Faucet Company.*)

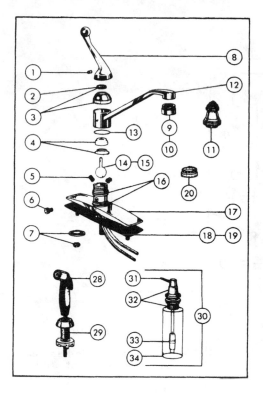

Figure 10.36 Single-handle kitchen faucet. (*Courtesy of Delta Faucet Company.*)

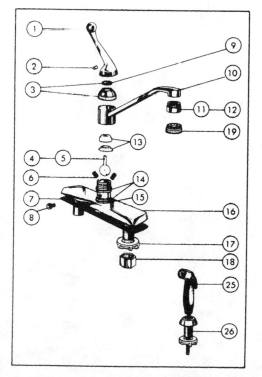

Figure 10.37 Single-handle kitchen faucet. (*Courtesy of Delta Faucet Company.*)

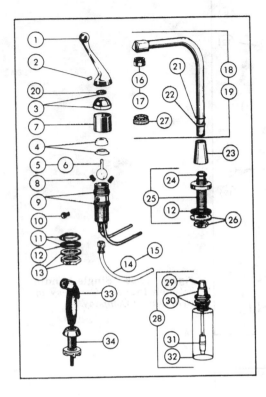

Figure 10.38 Single-handle, waterfall, single-mount utility faucet. (*Courtesy of Delta Faucet Company.*)

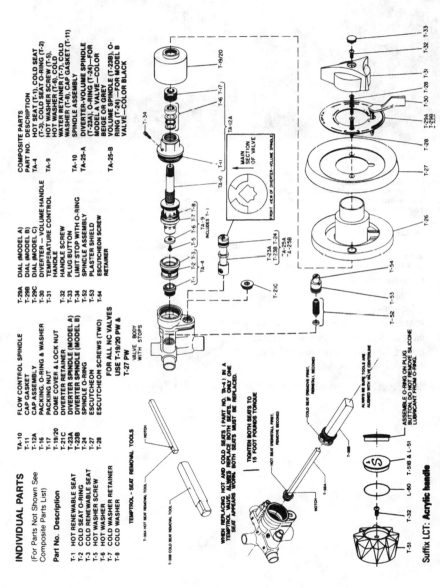

INDIVIDUAL PARTS

(For Parts Not Shown See Composite Parts List)

Part No.	Description
T-1	HOT RENEWABLE SEAT
T-2	COLD SEAT O-RING
T-3	COLD RENEWABLE SEAT
T-5	HOT WASHER SCREW
T-6	HOT WASHER
T-7	COLD WASHER RETAINER
T-8	COLD WASHER
TA-10	FLOW CONTROL SPINDLE
T-11	CAP GASKET
T-12A	CAP ASSEMBLY
T-16	PACKING, O-RING & WASHER
T-17	PACKING NUT
T-19/20	DOME COVER & LOCK NUT
T-21C	DIVERTER RETAINER
T-23A	DIVERTER SPINDLE (MODEL A)
T-23B	DIVERTER SPINDLE (MODEL B)
T-24	SPINDLE O-RING
T-27	ESCUTCHEON
T-28	ESCUTCHEON SCREWS (TWO)
T-29A	DIAL (MODEL A)
T-29B	DIAL (MODEL B)
T-29C	DIAL (MODEL C)
T-30	DIVERTER — VOLUME HANDLE
T-31	TEMPERATURE CONTROL HANDLE
T-32	HANDLE SCREW
T-33	PLUG BUTTON
T-34	LIMIT STOP WITH O-RING
T-52	SPINDLE ASSEMBLY
T-53	PLASTER SHIELD
T-54	ESCUTCHEON SCREW RETAINER

FOR ALL NC VALVES USE T-19/20 PW & T-27 PW

COMPOSITE PARTS

PART NO.	DESCRIPTION
TA-4	HOT SEAT (T-1), COLD SEAT (T-3), COLD SEAT O-RING (T-2)
TA-9	HOT WASHER SCREW (T-5), HOT WASHER (T-6), COLD WATER RETAINER (T-7), COLD WASHER (T-8), CAP GASKET (T-11) SPINDLE ASSEMBLY
TA-10	DIVERTER-VOLUME SPINDLE
TA-25-A	(T-23A), O-RING (T-24)—FOR MODEL A VALVE—COLOR BEIGE OR GREY
TA-25-B	VOLUME SPINDLE (T-23B), O-RING (T-24)—FOR MODEL B VALVE—COLOR BLACK

TEMPTROL—SEAT REMOVAL TOOLS

WHEN REPLACING HOT AND COLD SEATS (PART NO. TA-4) IN A TEMPTROL VALVE, ALWAYS REPLACE BOTH SEATS. IF ONLY ONE SEAT APPEARS WORN, BOTH SEATS MUST BE REPLACED.

TIGHTEN BOTH SEATS TO 15 FOOT POUNDS TORQUE

ASSEMBLE O-RING ON PLUG BUTTON, DO NOT REMOVE SILICONE LUBRICANT FROM O-RING.

ALWAYS BE SURE TOOLS ARE ALIGNED WITH VALVE CENTERLINE

COLD SEAT (REMOVE FIRST, REINSTALL SECOND)

HOT SEAT (REINSTALL FIRST, REMOVE SECOND)

Suffix LCT: Acrylic handle

Figure 10.39 Parts breakdown for a single-handle tub/shower valve. (*Courtesy of Symmons Industries, Inc.*)

Anti-Siphon Vacuum Breaker

NIDEL® Model 34H

NIDEL® Model 34HD

FOR NON-FREEZING AREAS ONLY

NIDEL® Model 34H anti-siphon vacuum breaker is designed to protect hose connections from contamination where temperatures do not get below freezing. Uses include laundry tubs, utility sinks, boiler room hose bibbs and outside hose bibbs in non-freezing areas.

FOR FREEZING AREAS

NIDEL® Model 34HD anti-siphon vacuum breaker is designed to protect hose connections from contamination where freezing temperatures are expected. Manual draining provides easy, sure drainage during cold weather. Uses include outside hose bibbs, wash racks, dairy barns and swimming pool areas.

SPECIFICATIONS:

APPROVALS — Approved under ASSE Standard 1011, Canadian Standards Association, listed by IAPMO® and accepted by U.S. Department of Health. For use downstream from the last shut-off valve. Not for continuous line pressure.

CONSTRUCTION — Brass and stainless steel construction.

DIAPHRAM — Flexible diaphram valve suitable for hot water.

ATTACHMENT SCREWS — Break-off.

PORTS — Dual safety design closes air ports during water flow.

INLET — ¾ inch standard female hose thread.

OUTLET — ¾ inch standard male hose thread.

FINISH — Plain brass. *Optional:* Chrome plated.

Figure 10.40   Antisiphon vacuum breaker. (*Courtesy of Woodford Mfg. Co.*)

Laboratory Vacuum Breakers
NIDEL® Model 38DF

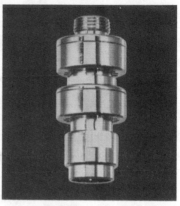

The Nidel Model 38DF vacuum breaker is designed for laboratory use to provide protection against back siphonage and back flow from aspirators or hoses left in contaminants.

DUPLEX CONSTRUCTION — Allows hose to be used above vacuum breaker.

CROSS CONNECTION PROTECTION — Guards against danger of cross connection contamination by protecting each faucet individually.

HIGH FLOW RATE — Sufficient flow rate to operate small aspirators.

EASILY ATTACHED — Gaskets provided to fit standard ⅜ inch laboratory faucets.

SPECIFICATIONS:

APPROVALS — Tested and approved by City of Los Angeles and City of Detroit Plumbing Laboratory. Accepted by U.S. Department of Health and many city and state health departments.

CONSTRUCTION — Brass and stainless steel.

INLET — ⅜ inch male pipe thread with gasket.

OUTLET — ⅜ inch straight female pipe thread.

DUPLEX VALVES — Dual-seal silicone diaphram valves for use in hot water.

PRESSURE REQUIRED — 7½ PSI minimum

FLOW RATE — 5¾ GPM @ 25 pounds differential pressure.

FINISH — Bright chrome.
Optional — Tin plated.
Optional — Special finishes and outlets available on request.

NOTE — Install only where drippage will go into sink or drain. Not recommended for distilled water.

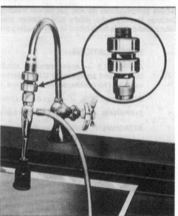

Figure 10.41   Laboratory vacuum breaker. (*Courtesy of Woodford Mfg. Co.*)

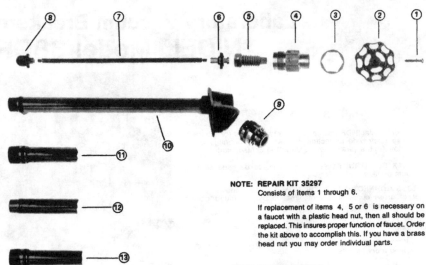

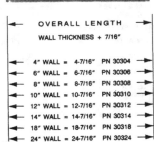

NOTE: REPAIR KIT 35297
Consists of items 1 through 6.

If replacement of items 4, 5 or 6 is necessary on a faucet with a plastic head nut, then all should be replaced. This insures proper function of faucet. Order the kit above to accomplish this. If you have a brass head nut you may order individual parts.

NOTE: 355XX ASSEMBLY
Consists of items 1 through 8 above. Order this assembly to replace all working parts of unit. However, you must furnish wall thickness.

OPERATING ROD
(As illustrated below)

MODEL 25 PARTS

ITEM	PART NO.	DESCRIPTION
1	30234	Handle Screw
2	30239	Wheel Handle - metal
3	30236	Drain Guard
4	30241	Head Nut - brass
5	30238	Stem Screw
6	35280	Drain Valve Assembly
1, 2, 3, 4, 5 and 6		Repair Kit (see above note)
7	303XX	Operating Rod
8	30230	Plunger Assembly
9	55057	Vacuum Breaker-brass
10	3546X	"CP" Inlet Casing Assembly (specify length)
11	3545X	"P" Inlet Casing Assembly (specify length)
12	3544X	"C" Inlet Casing Assembly (specify length)
13	3547X	"CP3" Inlet Casing Assembly (specify length)

OVERALL LENGTH

WALL THICKNESS + 7/16"

4" WALL = 4-7/16"	PN 30304
6" WALL = 6-7/16"	PN 30306
8" WALL = 8-7/16"	PN 30308
10" WALL = 10-7/16"	PN 30310
12" WALL = 12-7/16"	PN 30312
14" WALL = 14-7/16"	PN 30314
18" WALL = 18-7/16"	PN 30318
24" WALL = 24-7/16"	PN 30324

Figure 10.42 Parts breakdown for automatic-draining, freezeless wall faucets. (*Courtesy of Woodford Mfg. Co.*)

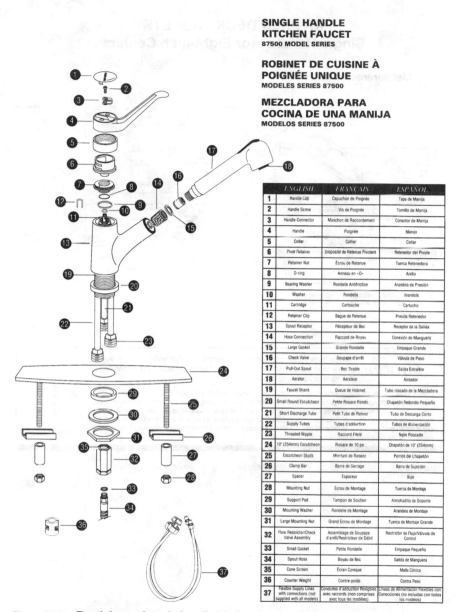

SINGLE HANDLE KITCHEN FAUCET
87500 MODEL SERIES

ROBINET DE CUISINE À POIGNÉE UNIQUE
MODELES SERIES 87500

MEZCLADORA PARA COCINA DE UNA MANIJA
MODELOS SERIES 87500

	ENGLISH	FRANÇAIS	ESPAÑOL
1	Handle Cap	Capuchon de Poignée	Tapa de Manija
2	Handle Screw	Vis de Poignée	Tornillo de Manija
3	Handle Connector	Manchon de Raccordement	Conector de Manija
4	Handle	Poignée	Manija
5	Collar	Collier	Collar
6	Pivot Retainer	Dispositif de Retenue Pivotant	Retenedor del Pivote
7	Retainer Nut	Écrou de Retenue	Tuerca Retenedora
8	O-ring	Anneau en «O»	Anillo
9	Bearing Washer	Rondelle Antifriction	Arandela de Presión
10	Washer	Rondelle	Arandela
11	Cartridge	Cartouche	Cartucho
12	Retainer Clip	Bague de Retenue	Presilla Retenedor
13	Spout Receptor	Récepteur de Bec	Receptor de la Salida
14	Hose Connection	Raccord de Boyau	Conexión de Manguera
15	Large Gasket	Grande Rondelle	Empaque Grande
16	Check Valve	Soupape d'arrêt	Válvula de Paso
17	Pull-Out Spout	Bec Tirable	Salida Extraíble
18	Aerator	Aérateur	Aireador
19	Faucet Shank	Queue de Robinet	Tubo roscado de la Mezcladora
20	Small Round Escutcheon	Petite Rosace Ronde	Chapetón Redondo Pequeño
21	Short Discharge Tube	Petit Tube de Renvoi	Tubo de Descarga Corto
22	Supply Tubes	Tubes d'adduction	Tubos de Alimentación
23	Threaded Nipple	Raccord Fileté	Niple Roscado
24	10" (254mm) Escutcheon	Rosace de 10 po	Chapetón de 10" (254mm)
25	Escutcheon Studs	Montant de Rosace	Pernos del Chapetón
26	Clamp Bar	Barre de Serrage	Barra de Sujeción
27	Spacer	Espaceur	Buje
28	Mounting Nut	Écrou de Montage	Tuerca de Montaje
29	Support Pad	Tampon de Soutien	Almohadilla de Soporte
30	Mounting Washer	Rondelle de Montage	Arandela de Montaje
31	Large Mounting Nut	Grand Écrou de Montage	Tuerca de Montaje Grande
32	Flow Restrictor/Check Valve Assembly	Assemblage de Soupape d'arrêt/Restricteur de Débit	Restrictor de Flujo/Válvula de Control
33	Small Gasket	Petite Rondelle	Empaque Pequeño
34	Spout Hose	Boyau de Bec	Salida de Manguera
35	Cone Screen	Écran Conique	Malla Cónica
36	Counter Weight	Contre-poids	Contra Peso
37	Flexible Supply Lines with connections (not supplied with all models)	Conduites d'adduction flexibles avec raccords (non comprises avec tous les modèles)	Líneas de Alimentación Flexibles con Conecciones (no incluidas con todos los modelos)

Figure 10.43 Breakdown of single-handle kitchen faucet. (*Courtesy of Moen, Inc.*)

KITCHEN DECK FAUCETS
Single-Handle - for Eight-Inch Centers

Measurements

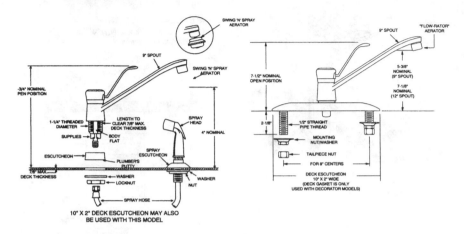

Equipped with Flow-Rator® Aerator or Swing-N-Spray Aerator

CAUTION: Always turn water off before disassembling the valve. Open valve handle to relieve water pressure and insure that complete water shut-off has been accomplished.

Before turning water on during either rough-in or trim-out, make sure that cartridge retainer clip is in place. The cartridge and retainer clip were properly installed and tested before leaving the factory. Although it is unlikely, it is nevertheless possible that through the handling of the valve by any number of persons the retainer clip may not be properly installed. This should be carefully checked at time of rough-in and trim-out. If the retainer clip is not properly installed, water pressure could force the cartridge out of the casting. Personal injury or water damage to the premises could result.

Figure 10.44 Single-handle kitchen deck faucet. (*Courtesy of Moen, Inc.*)

Disassembly

1. Turn "OFF" both hot and cold water supplies. Turn faucet on to relieve pressure and insure complete shut off.. Pull handle cap up and off (it snaps into place). Remove the handle screw (illustration 1).

2. Push the cartridge stem down and then lift and tilt handle lever and handle body off.

3. Unscrew retainer pivot nut.

4. Lift and twist spout off. The diverter can be removed at this point.

5. Pry out retainer clip with screwdriver (illustration 2).

6. Use cartridge twisting tool furnished with the new cartridge and rotate cartridge between 11 and 1 o'clock position.

7. Grasp cartridge stem with pliers. Pull cartridge out (illustration 3).

Reassembly

1. With cartridge stem UP, insert the cartridge with the ears aligned front to back (illustration 4).

2. Push the cartridge down by the ears (illustration 5) until flush with the top of the body.

3. Turn notched flat of cartridge stem toward front of sink. (Note: for cross piping installations, see instructions on back).

4. Replace the clip all the way (illustration 6). The diverter can be replaced at any time before the spout is installed.

5. Replace spout. Push down until it nearly touches the faucet escutcheon.

6. Screw on retainer pivot nut. DO NOT CROSS THREAD. Tighten firmly.

7. Press cartridge stem down. Holding handle UP, hook handle ring inside the handle body (illustration 7) into groove on the retainer pivot nut.

8. Swing handle back and forth until it drops down into place.

9. Replace handle screw. Tighten securely. Push handle cap down until it snaps into place.

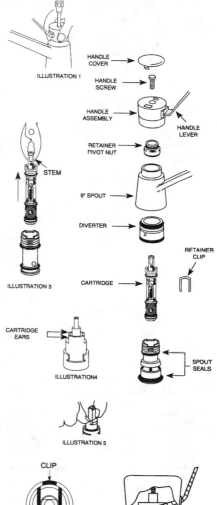

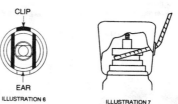

Figure 10.45   Disassembly and reassembly procedure for a cartridge-style, single-handle faucet. (*Courtesy of Moen, Inc.*)

RISER KITCHEN FAUCETS

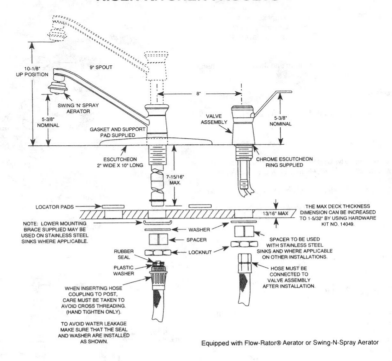

10-1/8"
UP POSITION

9" SPOUT

SWING 'N' SPRAY
AERATOR

5-3/8"
NOMINAL

GASKET AND SUPPORT
PAD SUPPLIED

8"

VALVE
ASSEMBLY

5-3/8"
NOMINAL

ESCUTCHEON
2" WIDE X 10" LONG

CHROME ESCUTCHEON
RING SUPPLIED

7-15/16"
MAX.

LOCATOR PADS

13/16" MAX

THE MAX DECK THICKNESS
DIMENSION CAN BE INCREASED
TO 1-5/32" BY USING HARDWARE
KIT NO. 14049.

NOTE: LOWER MOUNTING
BRACE SUPPLIED MAY BE
USED ON STAINLESS STEEL
SINKS WHERE APPLICABLE.

WASHER

SPACER

RUBBER
SEAL

LOCKNUT

SPACER TO BE USED
WITH STAINLESS STEEL
SINKS AND WHERE APPLICABLE
ON OTHER INSTALLATIONS.

PLASTIC
WASHER

HOSE MUST BE
CONNECTED TO
VALVE ASSEMBLY
AFTER INSTALLATION.

WHEN INSERTING HOSE
COUPLING TO POST,
CARE MUST BE TAKEN TO
AVOID CROSS THREADING.
(HAND TIGHTEN ONLY).

TO AVOID WATER LEAKAGE
MAKE SURE THAT THE SEAL
AND WASHER ARE INSTALLED
AS SHOWN.

Equipped with Flow-Rator® Aerator or Swing-N-Spray Aerator

CAUTION: **Always turn water off before disassembling the valve. Open valve handle to relieve water pressure and insure that complete water shut-off has been accomplished.**

Before turning water on during either rough-in or trim-out, make sure that cartridge retainer clip is in place. The cartridge and retainer clip were properly installed and tested before leaving the factory. Although it is unlikely, it is nevertheless possible that through the handling of the valve by any number of persons the retainer clip may not be properly installed. This should be carefully checked at time of rough-in and trim-out. If the retainer clip is not properly installed, water pressure could force the cartridge out of the casting. Personal injury or water damage to the premises could result.

CLIP

EAR

Figure 10.46   Riser-style kitchen faucet. (*Courtesy of Moen, Inc.*)

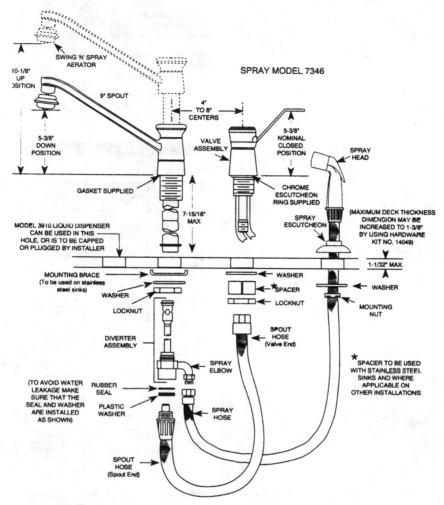

Figure 10.47   Breakdown of single-handle kitchen faucet with spray attachment. (*Courtesy of Moen, Inc.*)

TWO-HANDLE
TUB/SHOWER FAUCET
MODEL 82000 & 2500 SERIES.

ROBINET À POIGNÉE DOUBLE
POUR DOUCHE/BAIGNOIRE
MODÉLE SÉRIES 8200 ET 2500

MEZCLADORA PARA TINA/
REGADERA DE DOS MANIJAS
SERIES MODELOS 82000 y 2500

	ENGLISH	FRANÇAIS	ESPAÑOL
1	Shower Head	Pomme de Douche	Cebolleta
2	Shower Arm	Bras de Douche	Brazo de Regadera
3	Flange	Bride	Chapetón
4	Valve Body	Bâti de Soupape	Cuerpo de Válvula
5	Diverter	Dérivateur	Derivador
6	Union Adapter with Nut	Adaptateur de Raccordement avec Écrou	Adaptador de Unión con Tuerca
7	Cartridge	Cartouche	Cartucho
8	Cartridge Nut	Écrou de Cartouche	Tuerca del Cartucho
9	Escutcheon	Rosace	Chapetón
10	Stem Extension	Rallonge de Tige	Extensión de Vástago
11	Stem Extension Screw	Vis de Rallonge de Tige	Tornillo de la Extensión de Vástago
12	Handle Hub	Enjoliveur de Poignée	Centro de Manija
13	Optional Color Ring	Anneau de Couleur Facultatif	Anillo de Color Opcional
14	Lever Handle Insert (not included with all models)	Levier de Poignée Insérée (non compris avec tous les modèles)	Inserto de Manija de Palanca (no incluido con todos los modelos)
15	Handle Skirt	Jupe de Poignée	Faldón de Manija
16	Cross Handle Insert (not included with all models)	Pièce Insérée de Poignée Cruciforme (non compris avec tous les modèles)	Inserto de Manija en Cruz (no incluido con todos los modelos)
17	Cross Handle Screw	Vis de Poignée Cruciforme	Tornillo de Manija nen Cruz
18	Optional Color Ring	Anneau de Couleur Facultatif	Anillo de Color Opcional
19	Plug Button	Bouchon de Finition	Tapón
20	Cartridge Removal Tool	Outil de Démontage de Cartouche	Herramienta de Remoción
21	Tub Spout Escutcheon	Rosace de Bec de Baignoire	Chapetón de Salida de Tina
22	Tub Spout	Bec de Baignoire	Salida de Tina

Figure 10.48   Two-handle tub/shower faucet. (*Courtesy of Moen, Inc.*)

**SINGLE HANDLE
TUB/SHOWER FAUCET**
MODEL 3100 SERIES

**ROBINET À POIGNÉE UNIQUE
POUR DOUCHE/BAIGNOIRE**
SERIES MODÉLES 3100

**MEZCLADORA PARA TINA/
REGADERA DE UNA MANIJA**
SERIES MODELOS 3100

	ENGLISH	FRANÇAIS	ESPAÑOL
1	Shower Head	Pomme de Douche	Cebolleta
2	Shower Arm	Bras de Douche	Brazo de Regadera
3	Flange	Bride de Douche	Chapetón
4	Retainer Clip	Bague de Retenue	Pressilla Retenedora
5	Valve Body	Bâti de Soupape	Cuerpo de Válvula
6	Balance Spool	Tambour d'équilibre	Carrete de Balance
7	Cartridge	Cartouche	Cartucho
8	Check Stop	Soupape d'arrêt	Control de Parada
9	Adjustable Temperature Limit Stop	Régulateur de Température	Parada de límite de Temperatura Ajustable
10	Plaster Ground	Plaque Murale	Base para Pared
11	Escutcheon	Rosace	Chapetón

	ENGLISH	FRANÇAIS	ESPAÑOL
12	Escutcheon Screws	Vis de Rosace	Tornillos del Chapetón
13	Brake	Frein	Freno
14	Handle Stop	Arrêt de Poignée	Manija de Parada
15	Handle Stop Screw	Vis d'arrêt de Poignée	Retainer Tornillo de Manija
16	Handle Hub	Enjoliveur de Poignée	Centro de Manija
17	Lever Handle Insert	Vis de Blocage	Inserto de Manija de Palanca
18	Set Screw	Levier de Poignée Inséré	Tornillo de Sujeción
19	1/16" Hex Wrench	Clé Hexagonale de 1/16 po	Llave Hexagonal de 1/16"
20	Tub Spout Escutcheon	Rosace de Bec de Baignoire	Adjustable Temperature Limit Stop
21	Tub Spout	Rosace de Bec de Baignoire	Chapetón de Salida de la Tina
22	7/64" Hex Wrench	Clé Hexagonale de 7/64 po	Llave Hexagonal de 7/64"

Figure 10.49   Single-handle tub/shower faucet. (*Courtesy of Moen, Inc.*)

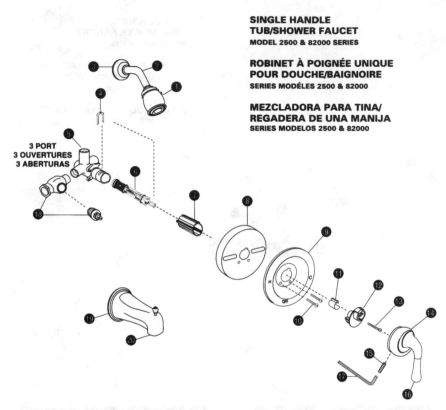

SINGLE HANDLE
TUB/SHOWER FAUCET
MODEL 2500 & 82000 SERIES

ROBINET À POIGNÉE UNIQUE
POUR DOUCHE/BAIGNOIRE
SERIES MODÉLES 2500 & 82000

MEZCLADORA PARA TINA/
REGADERA DE UNA MANIJA
SERIES MODELOS 2500 & 82000

3 PORT
3 OUVERTURES
3 ABERTURAS

	ENGLISH	FRANCAIS	ESPANOL
1	Shower Head	Pomme de Douche	Cebolleta
2	Shower Arm	Bras de Douche	Brazo de Regadera
3	Flange	Bride de Douche	Chapetón
4	Retainer Clip	Bague de Retenue	Presilla Retenedora
5	Valve Body	Bâti de Soupape	Cuerpo de Válvula
6	Cartridge	Cartouche	Cartucho
7	Stop Tube	Tube d'Arrêt	Tubo de Parada
8	Plaster Ground	Plaque Murale	Base para Pared
9	Escutcheon	Rosace	Chapetón
10	Escutcheon Screws	Vis de Rosace	Tornillos del Chapetón

	ENGLISH	FRANCAIS	ESPANOL
11	Brake	Frein	Freno
12	Handle Adapter	Adaptateur de Poignée	Adaptador de Manija
13	Handle Screw	Vis de Poignée	Tornillo de Manija
14	Handle Hub	Enjoliveur de Poignée	Centro de Manija
15	Set Screw	Vis de Blocage	Tornillo de Sujeción
16	Lever Handle Insert	Levier de Poignée	Inserto de Manija de Palanca
17	1/16" Hex Wrench	Clé Hexagonale de 1/16 po	Llave Hexagonal de 1/16"
18	Check Valve	Soupape d'Arrêt	Válvula de Control
19	Tub Spout Escutcheon	Rosace de Bec de Baignoire	Chapetón de Salida de la Tina
20	Tub Spout	Bec de Baignoire	Salida de la Tina

Figure 10.50 Single-handle tub/shower faucet. (*Courtesy of Moen, Inc.*)

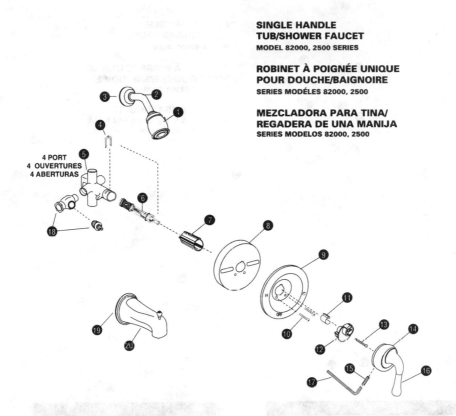

**SINGLE HANDLE
TUB/SHOWER FAUCET**
MODEL 82000, 2500 SERIES

**ROBINET À POIGNÉE UNIQUE
POUR DOUCHE/BAIGNOIRE**
SERIES MODÉLES 82000, 2500

**MEZCLADORA PARA TINA/
REGADERA DE UNA MANIJA**
SERIES MODELOS 82000, 2500

4 PORT
4 OUVERTURES
4 ABERTURAS

	ENGLISH	FRANÇAIS	ESPAÑOL
1	Showerhead	Pomme de Douche	Cebolleta
2	Arm	Bras de Douche	Brazo
3	Flange	Bride de Douche	Chapetón
4	Retainer Clip	Bague de Retenue	Presilla Retenedora
5	Valve Body (copper to copper)	Bâti de Soupape (cuivre/cuivre)	Cuerpo de válvula (cobre a cobre)
6	Cartridge	Cartouche	Cartucho
7	Stop Tube	Tube d'arrêt	Tubo de Parada
8	Plaster Ground	Plaque Murale	Base para Pared
9	Escutcheon	Rosace	Chapetón
10	Escutcheon Screws	Vis de Rosace	Tornillos de Chapetón

	ENGLISH	FRANÇAIS	ESPAÑOL
11	Brake	Frein	Freno
12	Handle Stop	Arrêt de Poignée	Parada de Manija
13	Handle Stop Screw	Vis d'arrêt	Tornillo de Parada
14	Handle Hub	Enjoliveur de Poignée	Centro de Manija
15	Set Screw	Vis de Blocage	Tornillo de Sujeción
16	Lever Handle Insert	Levier de Poignée	Inserto de Manija de Palanca
17	1/16" Hex Wrench	Clé Hexagonale de 1/16 po	Llave Hexagonal de 1/16"
18	Stop Valve	Soupape d'arrêt en Caoutchouc	Válvula de Parada Caucho
19	Tub Spout Escutcheon	Rosace de Bec de Baignoire	Chapetón de Salida de la Tina
20	Tub Spout	Bec de Baignoire	Salida de la Tina

Figure 10.51 Single-handle tub/shower faucet. (*Courtesy of Moen, Inc.*)

**SINGLE HANDLE
TUB/SHOWER FAUCET**
MODEL 82000, 2300

**ROBINET À POIGNÉE UNIQUE
POUR DOUCHE/BAIGNOIRE**
MODÉLES 82000, 2300

**MEZCLADORA PARA TINA/
REGADERA DE UNA MANIJA**
MODELOS SERIES 82000, 2300

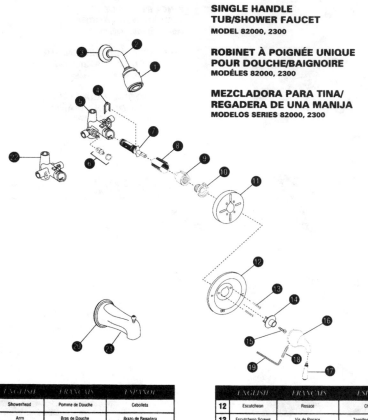

	ENGLISH	FRANÇAIS	ESPAÑOL
1	Showerhead	Pomme de Douche	Cebolleta
2	Arm	Bras de Douche	Brazo de Regadera
3	Flange	Bride de Douche	Chapetón
4	Retainer Clip	Bague de Retenue	Presilla Retenedora
5	Valve Body (copper to copper)	Bâti de Soupape (cuivre/cuivre)	Cuerpo de Válvula (cobre a cobre)
6	Rubber Stop Valve	Soupape d'arrêt en Caoutchouc	Válvula de Parada Caucho
7	Cartridge	Cartouche	Cartucho
8	Stop Tube	Tube de arrêt	Tubo de Parada
9	Key Stop	Clé d'arrêt	Candado de Parada
10	Temperature Limit Stop	Régulateur de Température	Parada de límite de Temperatura
11	Plaster Ground	Plaque Murale	Base para Pared

	ENGLISH	FRANÇAIS	ESPAÑOL
12	Escutcheon	Rosace	Chapetón
13	Escutcheon Screws	Vis de Rosace	Tornillos de Chapetón
14	Handle Stop	Arrêt de Poignée	Parada de Manija
15	Handle Stop Screw	Vis de Poignée	Tornillo Parada de Manija
16	Handle Hub	Enjoliveur de Poignée	Centro de Manija
17	Lever Handle Insert	Vis de Blocage	Inserto de Manija de Palanca
18	Set Screw	Levier de Poignée Inséré	Tornillo de Manija
19	1/16" Hex Wrench	Rosace de Bec de Baignoire	Llave Hexagonal de 1/16"
20	Tub Spout Escutcheon	Bec de Baignoire	Chapetón de Salida de la Tina
21	Tub Spout	Soupape de Bâti (tuyauterie d'adduction de fer)	Salida de la Tina
22	Valve Body (iron pipe supplies)	Clé Hexagonale de 1/16 po	Cuerpo de la Válvula (suministros de tubo de hierro)

Figure 10.52   Single-handle tub/shower faucet. (*Courtesy of Moen, Inc.*)

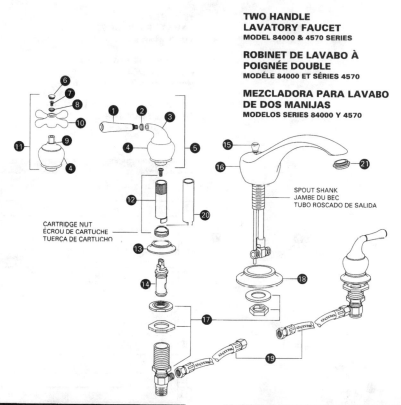

**TWO HANDLE
LAVATORY FAUCET**
MODEL 84000 & 4570 SERIES

**ROBINET DE LAVABO À
POIGNÉE DOUBLE**
MODÉLE 84000 ET SÉRIES 4570

**MEZCLADORA PARA LAVABO
DE DOS MANIJAS**
MODELOS SERIES 84000 Y 4570

SPOUT SHANK
JAMBE DU BEC
TUBO ROSCADO DE SALIDA

CARTRIDGE NUT
ÉCROU DE CARTUCHE
TUERCA DE CARTUCHO

	ENGLISH	FRANÇAIS	ESPAÑOL
1	Lever Handle Insert (not included with all models)	Levier de poignée inséré (non compris avec tous les modèls)	Inserto de Manija de Palanca (no incluido con todos los modelos)
2	Color Ring	Anneau de Couleur	Anillo de Color
3	Elbow	Coude	Codo
4	Handle Hub	Enjoliveur de Poignée	Centro de Manija
5	Lever Handle Assembly	Assemblage du Levier de Poignée	Ensamble de Manija de Palanca
6	Plug Button	Bouchon de finition	Bouchon de finition
7	Handle Screw	Vis de poignée	Vis de poignée
8	Color Ring	Anneau de Couleur	Anillo de Color
9	Handle Skirt	Jupe de Poignée	Faldón de Manija
10	Cross Handle Insert (not included on all models)	Poignée Cruciforme Insérée (non compris avec tous les modèles)	Inserto de Manija en Cruz (no incluido con todos los modelos)
11	Cross Handle Assembly	Assemblage de Poignée Cruciforme	Ensamble de Manija en Cruz

	ENGLISH	FRANÇAIS	ESPAÑOL
12	Stem Extension Kit (hot "red" and cold "blue")	Trousse de Rallonge de Tige (chaud-rouge/froid-bleu)	Juego de Extensión del Vástago (caliente "rojo" y frío "azul")
13	Handle Escutcheon	Rosace de Poignée	Escudo de Manija
14	Cartridge	Cartouche	Cartucho
15	Lift Rod & Knob	Bouton et Tige de Levée	Varilla Elevadora y Perilla
16	Spout Assembly	Assemblage du Bec	Ensamble de Salida
17	Hardware	Quincaillerie de Montage	Accesorios de Montaje
18	Spout Escutcheon	Rosace de Bec	Chapetón de Manija
19	Supply Hose	Boyau d'adduction	Manguera de Alimentación
20	Cartridge Removal Tool	Outil de Démontage de Cartouche	Herramienta de Remoción de Cartucho
21	Aerator	Aérateur	Aireador

Figure 10.53 Two-handle, three-piece lavatory faucet. (*Courtesy of Moen, Inc.*)

**TWO HANDLE
LAVATORY FAUCET**
MODEL 84000 & 4500 SERIES

**ROBINET DE LAVABO À
POIGNÉE DOUBLE**
MODÈLE 84000 ET SÉRIES 4500

**MEZCLADORA PARA LAVABO
DE DOS MANIJAS**
MODELO SERIES 84000 Y 4500

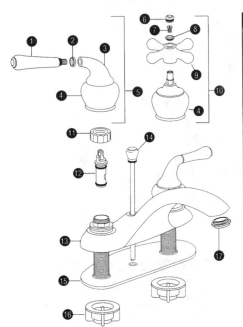

	ENGLISH	FRANÇAIS	ESPAÑOL
1	Lever Handle Insert (not included on all models)	Levier de Poignée Inséré (non compris avec tous les modèles)	Inserto de Manija de Palanca (no incluido con todos los modelos)
2	Color Ring	Anneau de Couleur	Anillo de color
3	Elbow	Coude	Codo
4	Handle Hub	Enjoliveur de Poignée	Centro de Manija
5	Lever Handle Assembly	Assemblage du Levier de Poignée	Ensamble de Manija de Palanca
6	Plug Button	Capuchon	Tapón
7	Handle Screw	Cartouche	Tornillo de Manija
8	Color Ring	Anneau de Couleur	Anillo de Color
9	Cross Handle Insert (not included on all models)	Poignée Cruciforme Insérée (non compris avec tous les modèles)	Inserto de Manija en Cruz (no incluido con todos los modelos)
10	Cross Handle Assembly	Assemblage de Poignée cruciforme	Ensamble de Manija en Cruz
11	Cartridge Nut	Écrou de Cartouche	Tuerca de Cartucho
12	Cartridge	Cartouche	Cartucho
13	Escutcheon	Rosace	Chapetón
14	Lift Rod & Knob	Tige de levée et bouton	Varilla Elevadora y Perilla
15	Deck Gasket	Support de Comptoir	Empaque de Cubierta
16	Mounting Nut	Écrou de Montage	Tuerca de Montaje
17	Aerator	Aérateur	Aireador

Figure 10.54   Two-handle lavatory faucet. (*Courtesy of Moen, Inc.*)

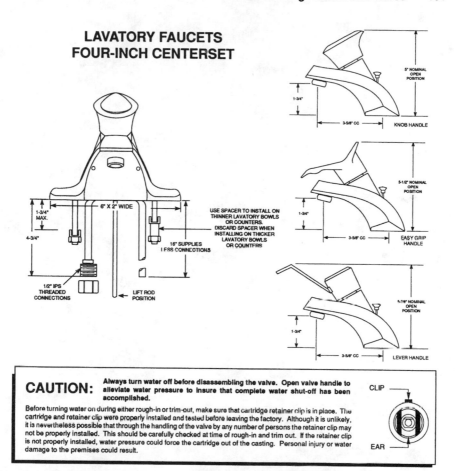

Figure 10.55   Single-handle lavatory faucet. (*Courtesy of Moen, Inc.*)

Chapter

11

Troubleshooting Plumbing Appliances

There are not many appliances that plumbers must work with, but there are a few. In some cases the appliances may only be connected to waste or water lines by plumbers. Other situations, such as those involving water heaters, require plumbers to have a good working knowledge of the appliance or device. Dishwashers, garbage disposers, and water heaters are the three major plumbing appliances most plumbers get involved with. Ice-makers are another appliance that plumbers sometimes work with.

In terms of plumbing connections, none of these appliances is very difficult to troubleshoot. However, when it comes to the inner parts of these same appliances, troubleshooting can become more difficult.

This chapter is going to go over all of the appliances mentioned above and cover the bases on effective troubleshooting procedures for each of them. We will start in the kitchen and work our way into the mechanical room.

Garbage Disposers

Garbage disposers are a common appliance in many homes. While plumbers rarely work on the motors of disposers, they are often the first service people called when a disposer is acting up. Problems with disposers can fall into three primary categories: plumbing, electrical, and appliance repair.

Our primary function in this chapter is to distinguish the causes of particular problems. While you, as a plumber, may not perform electrical work, you may be required to advise the property owner to call an electrician. Plumbers must be able to correct plumbing problems and refer other types of problems to appropriate repair people. As an example of knowing when to refer a customer to another type of service person, let's consider the following scenario.

A homeowner calls you to come out and repair a garbage disposer that is not working properly. When you arrive, you find that the disposer will cut on, but that it almost immediately trips out the reset button and goes off. After testing the disposer a few times, you are convinced the problem is most likely in the electrical wiring. Unless you are a licensed electrician, you should advise the homeowner of your opinion and suggest that perhaps an electrician should be called.

Under these conditions, you have fulfilled your role as a plumber. You responded to the call and did the initial troubleshooting. Seeing that the problem was not plumbing related, you gave your opinion and suggestions. That is about all that anyone could expect of you. I bring up the referral issue here because there are many times when working with appliances that the trouble is not in the domain of plumbers.

Now then, what can go wrong with a garbage disposer? Well, quite a bit, actually. To expand on this, let's take the probable troubleshooting phases on a one-at-a-time basis. We will do this in the form of stories about actual service calls.

In our first story, you have been called to a home to fix a garbage disposer that will not work. You arrive and turn on the garbage disposer. It makes a loud whirring sound, but doesn't dispose of its contents. What do you think is wrong with the disposer? If you said it's jammed, you're right. Now, how will you fix it?

Disposers that are suffering from jammed impeller blades can be fixed in one of three ways, depending on the type of disposer you are working with.

Expensive disposers often have a switch that allows the rotation of the impellers to be reversed, thereby freeing them from whatever is jamming them. This type of repair, when done by a plumber, often makes homeowners feel foolish and many plumbers can't look the customers in the face when they present their bill. As embarrassing as this situation can be for both parties, it is often simply a matter of turning on the reversing switch to solve the problem.

Many models of disposers are not equipped with reversing switches. A good number of these lower-grade disposers are, however, equipped with a special socket that allows the flywheel to be rotated. These sockets are usually hex shaped and located on the bottom of the unit. Wrenches for the socket are supplied with the disposers. A regular Allen wrench will get the job done if the factory-supplied wrench is not available. Once you have a wrench in the socket, all you have to do is turn the flywheel with the wrench to free the jam.

Another common method for freeing jammed impellers in disposers involves the use of a broom handle or similar tool. The broom handle is placed into the mouth of the disposer and lodged against one of the impeller blades. Then pressure is exerted on the blade as the handle is pried back to free the jam. The disposer should be turned off for all of these procedures, except for the use of a reversing switch.

In our second service call, you are called to a house where a disposer will not work. You turn on the switch and nothing happens. What would you do?

The first logical step to take is pushing the reset button that is located on the bottom of the disposer. Turn the disposer off and depress the reset button. Then turn the switch on and see if the disposer works. If it runs, test it several times to make sure it does not continue to trip the safety control.

What will you do if the disposer still fails to operate after using the reset button? Check to see if the impellers are jammed. For obvious reasons, never put your fingers or hands into the mouth of the disposer. Try turning the flywheel with a wrench or some long tool, like a broom handle. After rotating the flywheel, turn the disposer back on and see if it works.

If it is still dead, check the electrical panel to see if a fuse has blown or a circuit breaker has tripped. You'd be surprised how often this is the cause of the problem. Many homeowners, and commercial customers for that matter, never think to check their electrical panels before calling for help.

If the power to the disposer is on and it still won't run, you have two options. You can tell the customer the problem is not a plumbing problem and suggest that they call an electrician or appliance repair person. If you want to dig deeper, you can remove the wiring and test it to see that power is coming to the appliance. If full power is at the wires, the problem is likely to be with the disposer's motor.

Our next service call puts you in the position of troubleshooting foul odors being emitted from a disposer. The homeowner has used the disposer to gobble up everything from fish to fowl, and now the odors are permeating the house.

You arrive on the job and notice the smells as soon as you walk through the door. The odor is not that of sewer gas, but it certainly is unpleasant. What are you going to do first?

Are you thinking you should check the trap first? If you are, you're on the right track, sort of. If the trap is not maintaining its water seal, odors could be rising from the drain pipe. However, since the sink and disposer are being used regularly, it is unlikely the trap has lost its seal. Unless a vent has become obstructed and is allowing a siphonic action of the trap, there should be a good water seal. To be safe, you check the trap, and it is full of water, so the smells are not coming from the drain pipe. At this point you could tell the homeowner that the odor is coming from the appliance, not the plumbing and leave. This would be an acceptable course of action, but there is a better one available to you, if you know what to say. Do you?

Garbage disposers do sometimes become smelly beyond normal comfort zones. When this happens, there is a simple way to beat the smells. Fill the disposer about two-thirds full with ice cubes and cut it on. Run cold water into the disposer to flush it out. After the ice cubes have been ground up, slice a lemon in half and put it in the disposer. Cut the disposer on and allow it to devour the lemon. This home remedy works on almost any type of tough odor coming from disposers.

If a homeowner asks you whether it is best to flush a disposer with cold water or hot water, what would you say? Many people, plumbers included, would say hot water. In fact, I used to think hot water was the best to use when flushing a disposer. I found out, however, that cold water should be used. Why cold water? Cold water is said to congeal grease and carry it down the drain better than hot water. As I understand it, there is a higher likelihood of grease being caught on the sides of drainage piping if hot water is used.

Pure plumbing problems with disposers

Let's take a few moments to look at the pure plumbing problems often encountered with disposers. One of the most common is water leaking under the drain flange in the sink bowl. This is caused by a lack of putty under the flange. If this type of leak occurs on the back side of a disposer, the water from the leak can give the appearance of coming from somewhere else, especially from the trap. If you have a call where water is collecting beneath a sink bowl that is equipped with a disposer, check the drain flange carefully.

The lock-ring that holds a disposer in place can become loose, allowing water to leak. These occasions are rare, but they do sometimes happen.

Occasionally, the gasket between the disposer and the drain ell will leak. The leak could be a result of a bad washer or of the connection not being tight. A few quick turns of the lock-screw with a screwdriver will tell you if the problem is a loose connection or a bad washer.

It is also not uncommon for the slip nut that holds the discharge end of the drain ell in a trap to leak. This usually requires nothing more than tightening the slip nut, but a new slip-nut washer may be needed.

The mystery leak of a disposer

I'd like to tell you a story about the mystery leak of a disposer that perplexed two of my plumbers before the problem was solved. This situation happened a few years ago, and it is the only time I've ever heard of such a problem arising. Since this is a rare and unusual occurrence, I feel it warrants explanation.

My company received a call from a homeowner who was complaining of having water under her kitchen sink. A plumber was dispatched to the call, but he couldn't find the source of the leak. The plumber was licensed, but young and inexperienced. When he couldn't figure the problem out, a more seasoned plumber was dispatched.

The experienced plumber went through all the typical troubleshooting steps. After his best effort, he couldn't find the leak. He gave up and called me. I was out on the road and not too far away, so I went over to the house.

When I arrived, there was solid evidence that something was leaking into the base cabinet, under the sink. I conferred with my plumber and then began to troubleshoot the job.

I could see all the water connections, and none of them were even damp. I filled the left bowl of the sink with water and emptied it. Nothing leaked. Then I filled the right bowl, the one where the disposer was attached. The top to the disposer was missing, so I stuffed a rag in the opening to allow me to fill the sink. This was one step neither of my plumbers had taken.

When the right bowl was nearly full, I removed the rag and turned on the disposer. Guess what happened? The kitchen cabinet began flooding with water. The look on my plumber's face was worth the time I had spent on the call. He was somewhere between embarrassed and amazed. And I must admit, I was surprised to see what was happening. Do you know where the water was coming from?

The water was blowing out of the fitting where a dishwasher hose can be connected to a disposer. Apparently, someone had knocked out the drain plug in the side outlet of the disposer and never connected anything to it. Nor did they cap it. When the disposer was cut on, it forced water out of the uncovered drain outlet.

We capped the hole in the side of the disposer and the problem never recurred. My plumbers made two mistakes in their troubleshooting. First of all, they never bothered to turn the disposer on. Secondly, they didn't fill the bowl with the disposer to capacity and drain it. All they had done was run water down the drain from the faucet.

Even though I had no idea the water was coming out of the drain outlet, I took all of the right troubleshooting steps. I was surprised at the outcome of my test, but I was successful in finding and fixing the problem.

What is the moral to this story? It is simple, really. Take all the proper steps, don't take anything for granted, and you will probably solve any problem successfully.

Ice-Makers

Ice-makers are another good example of a situation in which the problem may not be in the plumbing. When someone's ice-maker fails to function properly, it is not unusual for them to call a plumber. From a pure plumbing point of view, the only aspect of an ice-maker that should concern a plumber is the connection at the back of the refrigerator, the tubing, and the connection to the cold-water pipe. These aspects of ice-makers can be troublemakers, but the problems are often within the ice-maker unit. When the problem is with the unit itself, the troubleshooting and repair should fall into someone else's field of expertise. This is not to say that some plumbers won't work on ice-maker units, but the units are not specifically plumbing.

If you, as a plumber, go poking around with an ice-maker unit without the knowledge and experience required to fix it, you could be buying yourself a lot of trouble. Suppose your dabbling winds up breaking the ice-maker? Guess who is going to have to pay to have it repaired or replaced? All I'm trying to say is that you should restrict your work to include only what you are an expert at. If you are not familiar with the workings of ice-makers, limit your work to the plumbing connections and tubing. There is nothing wrong with telling a customer that the problem is not plumbing related when the trouble is within the appliance.

Ice-makers are an appliance that many homeowners believe plumbers should know how to fix. Most plumbers, however, don't work on ice-makers. They make the necessary plumbing connections for the ice-makers but may not work on the unit itself. There are a few pure plumbing problems that can be associated with ice-makers, but more times than not, the trouble will be in the appliance, not in the plumbing.

What types of problems can arise for plumbers from ice-makers? Not too many, really. The saddle valve can create problems, and the tubing can leak, but that's really about all the pure plumbing involved with an ice-maker. Let's begin our short study of ice-makers with the saddle valve.

Saddle valves don't normally give plumbers much trouble. If the valve is not working properly, it is inexpensive enough to just replace it. Probably the most common complaint about saddle valves is their leaking at the packing nut. This is a simple problem, and one that is easily corrected by tightening the packing nut.

Sometimes a saddle valve will leak where the saddle comes into contact with the water distribution pipe. This is usually caused by the mounting screws being too loose or possibly by the gasket material being bad. Either cause is easily corrected.

The connection where the ice-maker tubing is attached to the refrigerator is another place where leaks are common. If the refrigerator is moved, say for cleaning purposes, the compression joint can become loose and leak. Again, this normally requires nothing more than tightening the nut.

If the tubing becomes kinked, the water flow to the ice-maker can be reduced or even cut completely off. This can also happen if the refrigerator is moved for any reason.

Plastic tubing sometimes gets cut when a refrigerator is being moved, and I've heard stories of plastic tubing just bursting for no apparent reason, though I personally have never seen that happen. Obviously, if the tubing ruptures or is cut, a serious leak will ensue.

When you are asked to troubleshoot an ice-maker problem, your first step should be to check the saddle valve. Make sure the valve is turned on. I don't know how these valves get cut off, but I've found several over the years that were in the closed position.

If the valve is open, pull the refrigerator out so that you can check the tubing connection to the appliance. Be careful not to tear the kitchen floor when you move the refrigerator.

Loosen the nut on the connection at the back of the refrigerator. If you begin getting a good spray of water, you know the appliance is receiving the water it needs to work. When this is the case, your job, as a plumber, is complete. You have proved that the appliance is being supplied with water, and that is all plumbers should be expected to do.

If for some reason there isn't any water pressure at the back of the appliance, backtrack to the saddle valve. Loosen the connection nut on the tubing at the saddle valve. If you have water at the valve and none at the appliance, you obviously have a problem in the tubing, probably a kink or possibly some type of obstruction.

If there is no water at the connection between the tubing and the saddle valve, you've either got a closed valve or a bad one. Try opening the valve, and if that doesn't work, replace it.

There really isn't much else to say about ice-makers, so let's move on to dishwashers.

Dishwashers

Dishwashers have become almost standard equipment in modern homes and commercial kitchens. While the problems that often occur with dishwashers are frequently appliance related, many people turn to plumbers for help. There are some problems associated with dishwashers that are pure plumbing problems, and many that aren't. We are going to concentrate on the pure plumbing problems and touch on the appliance-related problems.

The dishwasher won't drain

If you respond to a call where the dishwasher won't drain, what are you going to check first? There are several possibilities for why a dishwasher will not drain. Let's look at each of them individually.

Trap obstructions. Trap obstructions are one of the first considerations when troubleshooting a dishwasher that will not drain. This is an easy defect to check for. Fill the kitchen sink bowl that uses the same drain as the dishwasher and release the water to see if the sink drains properly. If it does, you can rule out the trap. If it doesn't, disassemble the trap and investigate.

Drain-pipe obstructions. Any drain-pipe obstructions will be found when you test the trap serving a dishwasher. If the sink using the same drain pipe that is being used by the appliance will not drain, you can count on the problem being in the drainage system. After inspecting the trap, you will know if the problem is in the trap or the drain pipe. If the pipe is clogged, you can snake it to get everything back in working order.

Air-gap obstructions. Air-gap obstructions are rare, but they can prevent a dishwasher from draining properly. This type of condition will normally be noticed when water draining from the dishwasher spills out of the air gap. Running a piece of wire through the air gap should solve the problem.

Kinked drainage hoses. Kinked drainage hoses are another common cause of poor drainage from dishwashers. To check for this, you will have to remove the access panel on the front of the appliance. A visual inspection will tell you if the hose is crimped or not.

Plugged disposer drainage inlets. Plugged disposer drainage inlets are a surefire cause of water being unable to drain from a dishwasher. Some rookie plumbers, and a lot of do-it-yourselfers, don't know that a plug must be knocked out of a disposer drain inlet prior to connecting dishwasher drainage hosing to it.

I've been to more than a few jobs where dishwashers were flooding kitchens through the air gap because the knockout plug in a disposer hadn't been removed. To experienced plumbers, this is such a silly concept that it sometimes goes unchecked, but let me tell you, check it. Remember how I told you earlier not to take anything for granted? Well, this is one of those cases. Never assume the installer of the dishwasher did the job right.

Stopped-up strainer baskets. Stopped-up strainer baskets, on the inside of dishwashers, can also inhibit proper drainage. While this is not a pure plumbing problem, it is one you should check. If the strainer is clogged with grease, food, or other debris, a quick cleaning will get the appliance back in operation.

The water to the dishwasher won't cut off

What is likely to be the problem when you are faced with a dishwasher where the water won't cut off when filling the appliance? There are only two probable causes for this situation. Either the solenoid is bad or the inlet valve is malfunctioning. Neither of these fall into the pure plumbing category, but you should be aware of them and their symptoms.

If you choose to get involved with appliance repair, you may have to repair or replace the solenoid. If the problem is with the inlet valve, you will have to disassemble and clean it. You may even have to replace it.

No water

What's wrong when there is no water coming into a dishwasher? A common problem is a closed valve on the supply tubing. Check the supply valve and make sure it is in the open position. Assuming the valve is open, you may have a faulty solenoid, a blocked inlet screen, a crimped tubing, or on a long-shot, inadequate water pressure.

If the problem is in the solenoid, it will have to be repaired or replaced, but again, this is not pure plumbing work.

The screen in the water inlet valve may be clogged with mineral deposits. You can check this and clean the screen if necessary.

Low water pressure is so unlikely that it is hardly worth mentioning, but if you suspect the pressure in the plumbing system is too low, you can put a pressure gauge on a hose bibb to test it. As long as you have normal pressure, something in the range of 40 pounds per square inch, you should be all right.

Crimped tubing is not a likely cause when you are working with a dishwasher that has worked properly in the past, unless it was recently moved and reinstalled. Drop the front access panel and do a visual inspection of the tubing to be sure it is not crimped.

Isn't cleaning properly

What should you look for with a dishwasher that isn't cleaning dishes properly? There are five typical things to check. Water temperature is one of them. If the water being used in the dishwasher is not hot, it will not clean as well as it should. The water should be as hot as code allows. When possible, a temperature of 160 degrees F is desirable.

Hard water can make dishwashers appear to not be doing their jobs. The film and residue left by hard water often makes dishes appear to have not been cleaned well. If you suspect this to be the problem, test the water for hardness.

Water pressure has an affect on how well a dishwasher works. If the water pressure to the appliance is too low, the spray arm cannot do its job as well, and the water will not beat the dishes clean in the way it should. Low water pressure is rarely a problem, but it could be.

It is a good idea to check the spray arm to see that it is not jammed. Twirl it around by hand and make sure nothing is inhibiting its rotation.

While you are working with the spray arm, check its holes to see that they are not plugged up. If the holes are blocked, water cannot be distributed to clean the dishes properly. Holes that are blocked can be opened with a piece of wire.

Leaks

Leaks from dishwashers can come from several places. In addition to the expected plumbing leaks, water can leak past a faulty door seal. Inspect the door seal to see if it is worn, torn, or out of position.

Water leaks under dishwashers can come from the compression fitting at the dishwasher ell or from the threaded portion of the ell. It is also possible that the supply tubing may have a hole in it. These leaks can usually be found easily by removing the access panel and inspecting the areas with a flashlight.

Drainage leaks are also possible under dishwashers. If the hose clamps are loose, leaks are likely. In addition to loose clamps, the insert fitting on the discharge outlet of the dishwasher may become cracked or broken. This doesn't happen often, but it can occur.

Washing Machines

Washing machines don't present many troubleshooting challenges. There are, however, a few tips worth talking about.

Leaks around the hose connections of washing machines are usually caused by bad or missing hose washers. By turning off the water valves and removing the washing machine hoses, you can do a quick inspection of the washers. If you suspect they are causing a leak, replace them.

Have you ever been called out because a washing machine would not fill with water? If you haven't, you probably will at some time. The major cause of this problem is debris in the hose screens. The troubleshooting process is simple.

Cut off the water at the washing machine valves. Remove the hoses from the valves and place a bucket under the valves. Turn the valves on and see if you have good water pressure; you probably will. This tells you that the problem is somewhere in the hoses or in the machine itself.

Remove the hoses from the back of the washing machine. There will be cone-shaped screens in the inlet ports. Inspect the screens for debris. If the screens are blocked, clean or replace them. It is not uncommon for the screens to become clogged to the point that water pressure in the washing machine is either very low or nonexistent. This is a common problem that many inexperienced plumbers are baffled by.

While drainage leaks are not frequently a problem with washing machines, they can exist. If the drain hose is leaking, check the clamps where the hose connects to the back of the machine. Put the machine in a drain cycle and watch to see that the hose is not cracked. Cracks in rubber hose can be very difficult to see unless they are under water pressure.

Keep an eye on the washing machine drain receptor as the machine drains. If the standpipe is too short, water may be spilling out the top of the indirect waste pipe. This seems to happen most frequently when large amounts of detergent are used in the washing cycle.

Electric Water Heaters

Electric water heaters (Fig. 11.1) can present a number of problems. Their thermostats can go bad (Fig. 11.2), the tanks can leak, the heating elements can burn out (Fig. 11.3), relief valves (Fig. 11.4) can pop off, and all sorts of related trouble can come up.

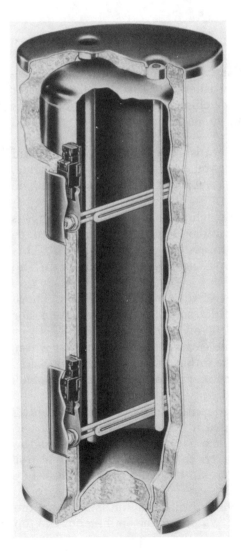

Figure 11.1 Cutaway of an electric water heater. (*Courtesy of A. O. Smith Water Products Co.*)

Figure 11.2 Upper thermostat for an electric water heater.

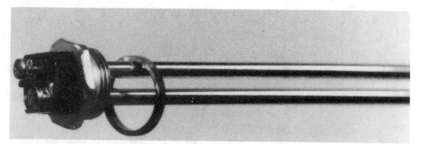

Figure 11.3 Screw-in heating element for an electric water heater.

Gas and electric water heaters are very different in the ways they work. They both do the same job, but they don't do it the same way. Let's begin our troubleshooting lesson with electric water heaters and close it with gas units.

I think it goes without saying, but beware of the electrical wires and current involved when working with electric water heaters. As you probably know, there is a lot of voltage running through the wires

Figure 11.4 Temperature and pressure relief valve for a water heater.

of a water heater. Once the access cover of an electric water heater is removed, you must be extremely careful not to touch exposed wires and connections.

Relief valves that pop off. Relief valves that pop off signal one of three problems: the relief valve is bad, the water heater is building excess pressure, or the heater is building excess temperature. The problem is usually just a defective relief valve. In such cases, replace the relief valve and monitor it to see that the new valve works properly. If the new valve releases a discharge, investigate for extreme temperature or pressure in the tank.

The temperature of water in the heating tank can be measured with a standard thermometer. Discharge a little water from the relief valve into a container and test its temperature. If it is too high for the rating of the temperature-and-pressure relief valve, check the thermostat settings on the water heater. Turn the heat settings down and test the water again after the new temperature settings have had time to work. If the reduction on the thermostat settings does not lower the temperature of the water in the tank, replacement of the tank is usually the best course of action.

If you suspect the water heater is under too much pressure, you can test the pressure with a standard pressure gauge. The easiest way to do this is to adapt the gauge to a hose-thread adapter and attach it to the drain at the bottom of the water heater. As long as the drain is not clogged, you can get an accurate pressure reading. You could also adapt the gauge to screw into the relief valve and test the pressure by opening the relief valve.

No hot water

When no hot water is being produced by an electric water heater, there are only a few things that need to be checked. The first thing to do is to check the electrical panel to see that the fuse or circuit breaker for the water heater is not blown or tripped. For water heaters that have their own disconnect boxes, check the disconnect lever to see that it is turned on.

Assuming the water heater is receiving adequate electrical power, check the thermostats to see that they are set at a reasonable heat setting. It is highly unlikely that anyone would turn them way down, but it is possible.

The most likely cause of this problem is a bad heating element, but a bad thermostat could also be at fault. Check the continuity and voltage of these devices with a meter to determine if they should be replaced.

Limited hot water

When a water heater is producing only a limited amount of hot water, the problem is probably with the lower heating element. These elements frequently become encrusted with mineral deposits that, in time, reduce the effectiveness of the element. Check the element with your meter and replace it if necessary.

When a thermostat is set too low, a water heater will not produce adequate hot water. Check the settings on the thermostats, and check the thermostats with your meter.

Water that is too hot

Sometimes complaints come in that the hot water being produced is too hot. This is normally a simple problem to solve. Turning the set-

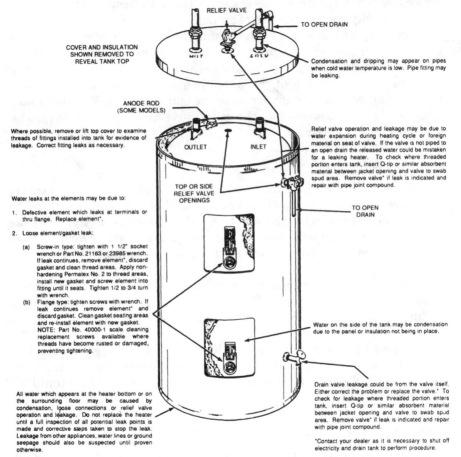

INSTRUCTIONS: USE THIS ILLUSTRATION AS A GUIDE WHEN CHECKING FOR SOURCES OF WATER LEAKAGE.
YOU OR YOUR DEALER MAY BE ABLE TO CORRECT WHAT APPEARS TO BE A PROBLEM.

RELIEF VALVE

TO OPEN DRAIN

COVER AND INSULATION
SHOWN REMOVED TO
REVEAL TANK TOP

HOT   COLD

Condensation and dripping may appear on pipes
when cold water temperature is low. Pipe fitting may
be leaking.

ANODE ROD
(SOME MODELS)

Where possible, remove or lift top cover to examine
threads of fittings installed into tank for evidence of
leakage. Correct fitting leaks as necessary.

OUTLET   INLET

Relief valve operation and leakage may be due to
water expansion during heating cycle or foreign
material on seat of valve. If the valve is not piped to
an open drain the released water could be mistaken
for a leaking heater. To check where threaded
portion enters tank, insert Q-tip or similar absorbent
material between jacket opening and valve to swab
spud area. Remove valve* if leak is indicated and
repair with pipe joint compound.

Water leaks at the elements may be due to:

1. Defective element which leaks at terminals or
thru flange. Replace element*.

2. Loose element/gasket leak:

(a) Screw-in type: tighten with 1 1/2" socket
wrench or Part No. 21163 or 23985 wrench.
If leak continues, remove element*, discard
gasket and clean thread areas. Apply non-
hardening Permatex No. 2 to thread areas,
install new gasket and screw element into
fitting until it seats. Tighten 1/2 to 3/4 turn
with wrench.
(b) Flange type: tighten screws with wrench. If
leak continues remove element* and
discard gasket. Clean gasket seating areas
and re-install element with new gasket.
NOTE: Part No. 40000-1 scale cleaning
replacement screws available where
threads have become rusted or damaged,
preventing tightening.

TOP OR SIDE
RELIEF VALVE
OPENINGS

TO OPEN
DRAIN

Water on the side of the tank may be condensation
due to the panel or insulation not being in place.

All water which appears at the heater bottom or on
the surrounding floor may be caused by
condensation, loose connections or relief valve
operation and leakage. Do not replace the heater
until a full inspection of all potential leak points is
made and corrective steps taken to stop the leak.
Leakage from other appliances, water lines or ground
seepage should also be suspected until proven
otherwise.

Drain valve leakage could be from the valve itself.
Either correct the problem or replace the valve.* To
check for leakage where threaded portion enters
tank, insert Q-tip or similar absorbent material
between jacket opening and valve to swab spud
area. Remove valve* if leak is indicated and repair
with pipe joint compound.

*Contact your dealer as it is necessary to shut off
electricity and drain tank to perform procedure.

Figure 11.5 Leakage checkpoints for a water heater. (*Courtesy of A. O. Smith Water Products Co.*)

tings down on the heater's thermostat will normally correct the situation. However, it is possible that the thermostat is defective. If it is, replace it.

Complaints of noise

Complaints of noise are sometimes made pertaining to water heaters. If a water heater is installed near habitable space, it is not uncommon for people to notice a rumbling in the heater. This often frightens people into calling a plumber. Do you know what causes the noise being made in the heater?

The noise is directly related to sediment that has accumulated in the water heater. Water becomes trapped between layers of the sediment, and when the water temperature reaches a certain point, the water explodes out of the layers. The miniexplosions can sound like a cracking or rumbling noise, and they should not be ignored. The steam explosions can be controlled by removing sediment from the water heater.

Sediment can be reduced in water heaters by opening the drain valve periodically. Ideally, water heaters should be drained monthly, but few are. If the buildup of sediment is extreme, a cleaning agent can be put into the water heater. The cleaning compound will reduce the scale buildup and the noise.

If you drain water from the heater and don't see any sediment, check the relief valve. It is possible the noise is coming from a steam buildup. When this is the case, you might need to replace the relief valve.

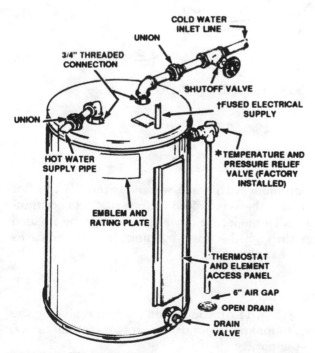

Figure 11.6   Electric water heater setup. (*Courtesy of A. O. Smith Water Products Co.*)

MODELS EDLJ/ELJF

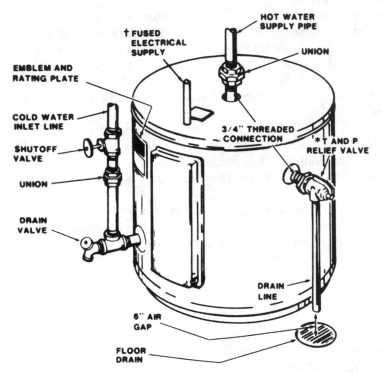

Figure 11.7 Electric water heater setup. (*Courtesy of A. O. Smith Water Products Co.*)

Rusty water

Rusty water can be a problem with some water heaters. If the hot water from fixtures is rusty, there is a good chance rust has accumulated in the water heater. In moderate cases the rust can be flushed out of the water heater through the drain opening. In extreme cases the entire heater should be replaced.

When the tank is leaking

When the tank is leaking (Fig. 11.5), you should plan on replacing the whole unit. It is possible to make temporary repairs for leaking tanks, but they are just that—temporary.

If you need to plug a hole temporarily, you can do it with a toggle

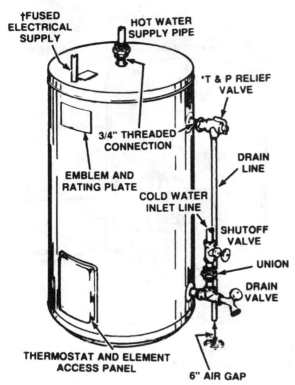

Figure 11.8 Electric water heater setup. (*Courtesy of A. O. Smith Water Products Co.*)

bolt and a rubber washer. Drain the water heater to a point below the leak. Drill a hole in the tank that will allow you to insert the toggle bolt. The toggle bolt should have a metal washer against its head and a rubber washer that will come into contact with the tank. Once the toggle penetrates the tank and spreads out, tighten the bolt until the rubber washer is compressed. This will slow down or stop the leak for a while, but don't expect the repair to last indefinitely. Once a heater starts to develop pinholes, it is time to replace it.

That pretty well covers the range of problems associated with electric water heaters. As long as you know the components you are working with, water heaters are not difficult to troubleshoot (Figs. 11.6 through 11.11 and Table 11.1). Now let's turn our attention to gas-fired heaters and see how the troubleshooting methods differ.

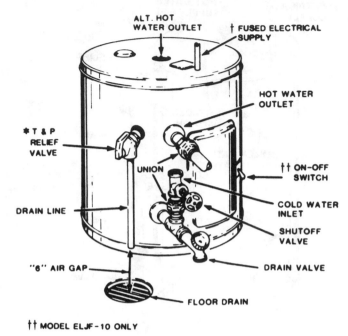

Figure 11.9   Electric water heater setup. (*Courtesy of A. O. Smith Water Products Co.*)

Gas-Fired Water Heaters

Gas-fired water heaters (Fig. 11.12) share some of the same trouble symptoms that electric water heaters produce. However, there are many differences between the two types of heaters (Fig. 11.13). Let's look at the same problems that we studied for electric water heaters and see how your job will differ when troubleshooting gas units.

Relief valves

Relief valves on gas heaters can be tested and treated the same as those used on electric heaters.

No hot water

When you are getting no hot water from a gas-fired water heater, the first thing to check is the pilot light. If the pilot light is not burning, check the gas valve to see that it is turned on. If the gas valve is on

MODELS
ECTT/ETTN/ESTT

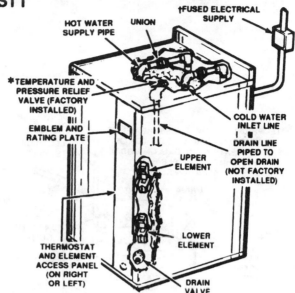

HOT WATER SUPPLY PIPE

UNION

†FUSED ELECTRICAL SUPPLY

＊TEMPERATURE AND PRESSURE RELIEF VALVE (FACTORY INSTALLED)

EMBLEM AND RATING PLATE

COLD WATER INLET LINE

DRAIN LINE PIPED TO OPEN DRAIN (NOT FACTORY INSTALLED)

UPPER ELEMENT

LOWER ELEMENT

THERMOSTAT AND ELEMENT ACCESS PANEL (ON RIGHT OR LEFT)

DRAIN VALVE

Figure 11.10 Tabletop-style water heater. (*Courtesy of A. O. Smith Water Products Co.*)

and the pilot light is not burning, you must try to relight the pilot. However, before doing this, make sure that is not an accumulation of trapped gas that will explode when you light a flame.

Cut the gas valve off and ventilate the area well. When you are sure it is safe to light the pilot, turn the gas valve on and light the pilot. If the pilot will not light, make sure there is gas coming through the piping. You should be able to hear or smell it.

The thermostat may be turned off or defective. If the thermostat is set in the proper position but won't function, replace it.

The thermocoupling could also be bad. If the pilot light lights but continues to go out, replace the thermocoupling.

If none of these methods prove fruitful, check the dip tube. It should be installed in the cold-water side of the tank, and it is possible that it was put in the hot-water side by mistake.

Limited hot water

When a gas-fired water heater produces only limited hot water, the problem is usually with the gas control. Check the setting to see that

MODELS EEC/EEH/EES/EEST/PEC/PED/PEH/PEN

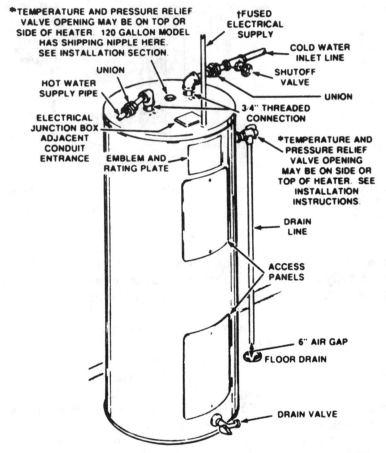

Figure 11.11 Electric water heater setup. (*Courtesy of A. O. Smith Water Products Co.*)

it is set high enough to produce a satisfactory supply of hot water. If it is set properly but failing to work, replace it.

The dip tube could be responsible for the production of limited hot

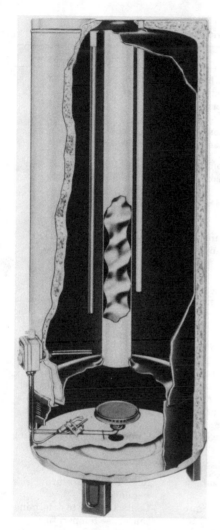

Figure 11.12 Cutaway of a gas water heater. (*Courtesy of A. O. Smith Water Products Co.*)

water. If it is installed in the hot-water side or is broken, you will have to either replace it or move it to the cold-water side of the tank.

Water that is too hot

When the water from a gas-fired water heater is too hot, the gas control valve is either bad or set too high. Check the setting and adjust it to a lower level. If that doesn't solve the problem, replace the valve assembly.

TABLE 11.1   Troubleshooting Guide for Electric Water Heaters

Complaint	Cause	Solution
Water leaks (See leakage checkpoints on page 7.)	Improperly sealed hot or cold supply connections, relief valve, or thermostat threads	Tighten threaded connections
	Leakage from other appliances or water lines	Inspect other appliances near water heater
Leaking t & p valve	Thermal expansion in closed water system	Install thermal expansion tank. (do not plug t & p valve)
	Improperly seated valve	Check relief valve for proper operation
Hot water odors (Caution: unauthorized removal of the anode(s) will void the warranty. For further information, contact your dealer.)	High sulfate or mineral content in water supply	Drain and flush heater thoroughly; refill
	Bacteria in water supply	Chlorinate water supply
Not enough or no hot water	Power supply to heater is not on	Turn disconnect switch on or contact electrician
	Thermostat set too low	Refer to temperature regulation
	Heater undersized	Reduce hot water use
	Incoming water is unusually cold water (winter)	Allow more time for heater to reheat
	Leaking hot water from pipes or fixtures	Check and repair leaks
	High-temperature limit switch activated	Contact dealer to determine cause; refer to temperature regulation
Water too hot	Thermostat set too high	Refer to temperature regulation
	High-temperature limit switch activated	Contact dealer to determine cause; see temperature regulation
Water heater sounds	Scale accumulation on elements	Contact dealer to clean or replace elements
	Sediment buildup in tank bottom	Drain & flush thoroughly

SOURCE: A. O. Smith Water Products Co.

WATER HEATING CYCLE
(GAS AND ELECTRIC POWER ARE ON, "OFF/ON" SWITCH IS ON)

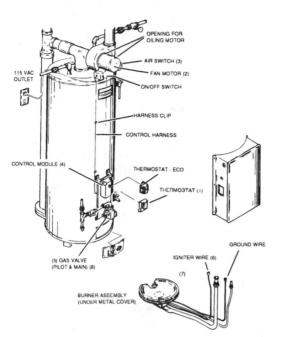

1) THERMOSTAT (1) CALLS FOR HEAT
   A) THERMOSTAT SENSES NEED FOR HEATING WATER
   B) CONTACTS CLOSE IN THERMOSTAT, POWER FLOWS TO BLOWER MOTOR (2)

2) BLOWER MOTOR ROTATES BLOWER WHEEL
   A) BLOWER WHEEL SPEED INCREASES
   B) WHEEL SPEED INCREASES, AIR PRESSURE SWITCH (3) CONTACTS CLOSE.

3) AIR PRESSURE SWITCH CONTACTS CLOSE
   A) 24 VAC FLOWS TO IGNITION CONTROL MODULE (4)
   B) CONTROL MODULE STARTS IGNITION SEQUENCE

4) 24 VAC FLOWS TO PILOT VALVE COIL (5)
   A) COIL OPENS PILOT VALVE
   B) GAS FLOWS TO PILOT ORIFICE

5) IGNITER (6) STARTS TO SPARK
   A) SPARK ACTION STARTS AT PILOT
   B) PILOT LIGHT IGNITES

6) MODULE (4) SENSES PILOT FLAME
   A) MODULE OPENS MAIN (8) GAS VALVE
   B) GAS FLOWS TO MAIN (7) BURNER

7) MAIN BURNER (7) IGNITES
   A) GAS FLOWS TO MAIN BURNER PORTS AND IS IGNITED

8) BURNER HEATS WATER
   A) BURNER HEATS WATER TO THERMOSTAT SETTING
   B) CONTACTS IN THERMOSTAT (1) OPEN, BURNER GOES OUT AND BLOWER SHUTS DOWN
   C) CYCLE IS COMPLETE

Figure 11.13   Heating cycle for a water heater. (*Courtesy of A. O. Smith Water Products Co.*)

Noise, rusty water, and leaks in tanks

Noise, rusty water, and leaks in tanks can be treated the same for gas-fired heaters as for electric water heaters. Refer to the instructions given earlier for electric water heaters.

Gas odors

Gas odors are sometimes noticed around gas-fired water heaters. This can be a dangerous situation. The problem is usually a leak in a fitting, pipe, or piece of tubing. Use soapy water on all places where a leak might occur to find the source of the smell. The soapy water will bubble when it is applied to the leaking location.

If all of the connections, pipe, and tubing check out to be okay, inspect the venting of the water heater. Inadequate or improper venting could cause gas smells to be trapped around the heater.

We have now concluded our look at troubleshooting plumbing appliances, and we are ready to move onto the troubleshooting of wastewater pumps.

12

Troubleshooting Waste-Water Pumps

Waste-water pumps come in all shapes and sizes. The jobs they do are as varied as the configurations in which they are made. Some waste-pumping systems are equipped with high-water alarms, and some are so simple that almost anyone can install and operate them.

When you are called out to troubleshoot a waste-water pumping station, you could be faced with a seized impeller, a float ball that has stuck in one position, a check valve that refuses to close, a gate valve that has been closed, or any number of other possible problems.

The troubleshooting of waste-water pumps and the systems surrounding them is not particularly complicated. From a pure plumbing point of view, there are relatively few things that will impede the smooth operation of such systems. As with all troubleshooting, a sensible plan of attack is usually all that is necessary to make short work of finding pump problems.

When a customer's waste-water pump fails to operate properly, the customer can be quite panicked. A homeowner who is suddenly startled by a pump-station alarm going off can be difficult to calm, and business owners who are watching water levels in their sump basins rise tend to be short tempered. For a plumber, this is not a good time to be too busy to soothe the customer or to respond quickly to the call for help.

The failure of a waste-water pump can bring a business to a complete stop, and a problem pump can cripple a household. Some plumbing systems rely on their pump stations to evacuate all waste from the system; others only depend on pumps for individual fixtures or special uses.

Most pumps used to handle waste water are of the submersible type. There are, however, a large number of pedestal sump pumps in operation. Depending upon the usage of the pump, it may be small enough to mount on the drain of a laundry tub or large enough to require an extra set of hands to remove it from its basin.

Pumps that have integral floats tend to be less troublesome than those with external floats. External float frequently get stuck in either the pump-on or the pump-off position. This is a simple problem to solve, but it is an annoying one to have.

Where are waste-water pumps used? They are used in both residential and commercial applications. They can be used to facilitate the draining of washing machines, laundry tubs, rainwater piping, or entire plumbing systems.

There are five types of waste-water pumps that we are going to examine in this chapter; they are submersible sump pumps, pedestal sump pumps, sewage ejector pumps, basin waste pumps, and grinder pumps. Let's begin our journey into waste-water pumps with a discussion of submersible sump pumps.

Submersible Sump Pumps

Submersible sump pumps generally work very well, but there are times when they fail to perform properly. The most obvious difference between various types of submersible sump pumps is the location of their floats. Some styles have their floats concealed within the pump housing (Fig. 12.1). Integral floats are desirable since they are unlikely to become stuck or jammed by foreign objects.

Some pumps have their floats mounted on a metal rod that controls the path of the float. The float operation on these pumps is not quite as reliable as it is for pumps with integral floats, but the dependability is still good.

A third style of pump has a float on an arm (Fig. 12.2). This type of float sometimes gets stuck against the side of the pump basin. If the pump is installed too close to the basin, the float is sure to cause problems.

A fourth style of pump uses an independent float, one that is not put on the pump. These floats are often attached to the discharge pipe of the pump, but some units are attached to a special clip that is mounted on the pump housing. The float cable provides potential for problems, with the float becoming stuck in an undesirable position.

The normal size of the discharge outlet on a sump pump is $1\frac{1}{4}$ inches. Sometimes the discharge line is run with flexible $1\frac{1}{4}$-inch pipe, and sometimes it is piped with $1\frac{1}{2}$-inch pipe. Larger sump pumps can

▼ Sump, Sewage, Effluent and Grinder Pumps

A.Y. McDonald Mfg. Co. offers a full line of Sump, Effluent, Ejector and Grinder Pumps for home and light industrial applications.

Sump Pumps - featuring high torque capacitor-start motors. Rustproof construction, clog proof vortex impeller, easy to service. Available in 1/4, 1/3, 1/2 and 1 1/2 H.P. models.

Sewage Ejectors and Effluent Pumps - capable of handling 3/4", 2" and 2 1/2" solids, available in 4/10 through 3 H.P. models for residential and light industrial applications.

Grinder Pumps - for disintegrating solids. For pressure sewer systems, restaurants, and industry. Replaces systems that need screening or occasionally clog. Available in a convenient 2 H.P. packaged system.

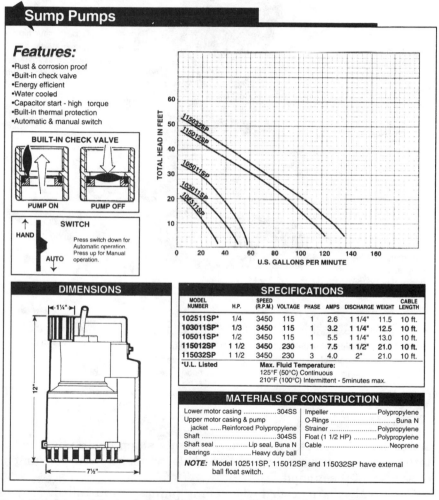

Figure 12.1 Internal-float, submersible sump pump. (*Courtesy of A. Y. McDonald Mfg. Co.*)

have discharge outlets up to 2 inches in diameter. The final discharge location for the drain of a sump pump can be hard to predict.

Storm water is not usually supposed to be piped into a sanitary

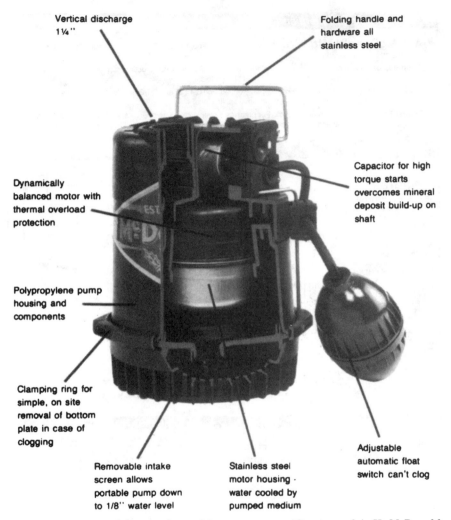

Vertical discharge
1¼"

Folding handle and
hardware all
stainless steel

Capacitor for high
torque starts
overcomes mineral
deposit build-up on
shaft

Dynamically
balanced motor with
thermal overload
protection

Polypropylene pump
housing and
components

Clamping ring for
simple, on site
removal of bottom
plate in case of
clogging

Adjustable
automatic float
switch can't clog

Removable intake
screen allows
portable pump down
to 1/8" water level

Stainless steel
motor housing ·
water cooled by
pumped medium

Figure 12.2   External-float, submersible sump pump. (*Courtesy of A. Y. McDonald Mfg. Co.*)

sewer, but it often is. The discharge from a sump pump might dump into a storm sewer or a gravel-lined hole or might even just run out on the top of the ground.

Check valves are needed when sump pumps are installed; without them, the pumps would run continuously. When the pump finishes its pumping cycle, water between the pump and the point where gravity drainage begins is trapped in the vertical pipe. Without a check valve, this water would run back into the sump and force the pump to cut back on. This on-and-off action would destroy the pump's motor. The check valves used for sump pumps may be installed in the discharge piping, or they may be built into the pump.

Some sump pumps are equipped with both manual and automatic operation switches. The automatic mode is the normal setting when the pump is installed. The voltage requirement for average sump pumps is 115 volts. Larger models require 230 volts.

The volume of water that can be pumped and the vertical distance that can be obtained from a sump pump depends on the size of the pump's motor.

A small sump pump, one with a ¼-horsepower motor, can pump approximately 33 gallons per minute with a vertical head of about 20 feet. The same type of pump with a ⅓-horsepower motor will move about 50 gallons of water per minute with a head of about 25 feet. A ½-horsepower motor on the pump will produce a volume of about 58 gallons per minute with a head of about 32 feet. A large sump pump, one with a 1½-horsepower motor, can deliver about 140 gallons per minute at a head of about 52 feet. Now that we have a good overview of what submersible sump pumps are, let's take a look at what types of problems are often associated with them.

No check valve

A sump pump with no check valve is going to cycle on and off frequently, if not constantly. If you have a sump pump that is cutting on and off too often, investigate the check valve. You may find that one was never installed, or you might discover that the check valve is stuck in the open position. In either case, the pump will run too often.

If there is no check valve present, you should install one. If the existing check valve is stuck in the open position, you may be able to repair it, or you may have to replace it.

Closed check valve

A closed check valve will prohibit the pump from emptying its sump. If a check valve becomes stuck in the closed position, water cannot move through the drain pipe. This type of problem usually calls for the replacement of the check valve, but there are times when the flapper can be manually opened and repaired.

Frozen drain pipes

Frozen drain pipes can prevent sump pumps from doing their jobs. It is not uncommon for the drain lines from sump pumps to be run in a less-than-professional manner. This is especially true when the drain is run with flexible hose-type pipe.

If the drain from a sump pump is run on top of the ground or in a shallow trench, it is subject to freezing during cold temperatures. If the pipe does not have a good grade on it or if the pipe loops downward in spots, freezing becomes more likely.

If you are faced with a pump that runs but won't pump water, it is usually a problem with the check valve, but it could be that the drain pipe has frozen.

You can confirm if the problem is with the pump by disconnecting the drain pipe and running the pump. With the pipe disconnected, you will be able to see if the pump is producing water.

Clogged strainers

Clogged strainers routinely cause sump pumps to fail. When you have a pump that is not pumping properly, pull the pump and inspect the strainer on the bottom of the inlet opening. It is not uncommon for the strainer to be blocked with leaves, paper, or all kinds of other strange objects that shouldn't be in a sump.

If the pump you are working on does have a clogged strainer, all you should have to do is clean it. Once the strainer is clear, the pump should work correctly.

Jammed impellers

Jammed impellers can stop a pump dead in its tracks. If a pebble or some other object finds its way into the impellers, the pump can stop pumping. The motor will run, but no water will be produced. When this is the case, you must unplug the pump, gain access to the impellers, and clear them.

Floats that are stuck

Floats that are stuck can cause two different types of problems. If a float becomes stuck at its upper level, the pump it is serving will not cut off. When the float sticks at its lower level, the pump will not cut on. Both of these problems are serious and require fast attention.

Floats can be jammed by objects floating in the sump, but they most often stick without apparent cause, except for when they rub against the sump and become stuck.

The vibrations that go through a sump pump when it runs are enough to make the pump walk across the sump pit. Unless the pump is piped into place with rigid pipe, it can creep across the sump until the float begins to hit the side of the container. This results in a float that sticks and a pump that doesn't work properly.

Unplugged electrical cables

Unplugged electrical cables account for some pump problems. This situation may seem too simple to even address, but sometimes pumps become unplugged, and they don't run well without electricity. When you have a pump that won't run at all, check to see that it is plugged in and that the fuses or circuit breakers are in good condition and in the proper positions.

Pedestal Sump Pumps

Pedestal sump pumps are very common in residential properties. These pumps are inexpensive and do a pretty good job of handling modest pumping demands. The major difference in a pedestal pump and a submersible pump is that the motor of a pedestal pump is above the water level. Other than this, a pedestal sump pump can be treated like a submersible pump for troubleshooting purposes.

Sewage Ejector Pumps

Sewage ejector pumps are ones nobody wants to have trouble with. When these pumps fail to do their jobs, working conditions can get messy. Properties that depend on ejector pumps to evacuate their sewage cannot afford to have their pumps fail.

Sewage ejector pumps (Fig. 12.3) are not exactly like sump pumps, but there are similarities. For example, the impellers on a sewer pump can become jammed, just like those in a sump pump. Check valves that are used for sewer pumps can fail in the same ways as described for sump pumps. There are, however, a number of differences between the installation and operation of a sewer pump and a sump pump.

Standard sewer ejectors have 2-inch discharge outlets and are capable of handling 2-inch solids. Some plumbers install effluent pumps for use in draining sewage sumps. This is a mistake. Effluent pumps typically have 2-inch discharge outlets, but they are only rated to handle ¾-inch solids. If an effluent pump is installed for use as a sewage ejector, the size of the solids that need to be pumped can create problems. When you are troubleshooting a sewage ejector, make sure that it is, in fact, a sewage ejector and not an effluent pump.

A sewer ejector with a ½-horsepower motor is usually all that is needed for most pump stations. These pumps can push about 175 gallons a minute up to a height of nearly 24 feet. These pumps are available in either 115 or 230 volts. The electrical cables supplied on sewage ejectors are typically about 15 feet long.

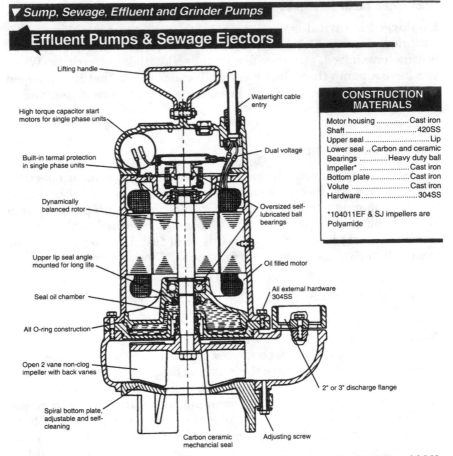

▼ *Sump, Sewage, Effluent and Grinder Pumps*

Effluent Pumps & Sewage Ejectors

Lifting handle

Watertight cable entry

High torque capacitor start motors for single phase units

Dual voltage

Built-in termal protection in single phase units

Dynamically balanced rotor

Oversized self-lubricated ball bearings

Upper lip seal angle mounted for long life

Oil filled motor

Seal oil chamber

All external hardware 304SS

All O-ring construction

Open 2 vane non-clog impeller with back vanes

Spiral bottom plate, adjustable and self-cleaning

2" or 3" discharge flange

Carbon ceramic mechancial seal

Adjusting screw

CONSTRUCTION MATERIALS	
Motor housing	Cast iron
Shaft	420SS
Upper seal	Lip
Lower seal	Carbon and ceramic
Bearings	Heavy duty ball
Impeller*	Cast iron
Bottom plate	Cast iron
Volute	Cast iron
Hardware	304SS

*104011EF & SJ impellers are Polyamide

Figure 12.3 Cutaway of a submersible effluent pump. (*Courtesy of A. Y. McDonald Mfg. Co.*)

Sewage ejector pumps are available with 3-inch discharge flanges, but a 2-inch flange is more common. Depending on the brand of pump, the float can be a part of the pump, or it can be an independent unit that plugs in with the pump through a piggyback plug. Unfortunately, the piggyback plug type of float sometimes gets tangled or stuck on the side of the sump (Fig. 12.4). Let me give you a real-world example that I faced recently.

I was at my insurance agent's office setting up a new liability policy. As we talked, my agent asked me some questions about his sewage ejector system. He was complaining that his sewer pump was cutting on too often. I made an appointment with him to check it out.

When I got to my customer's house, I flushed the toilet that was connected to the sump. The toilet was a 1½-gallon model that should

COMPLETE BASIN

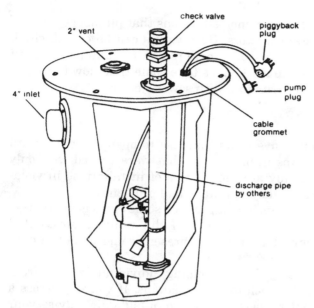

Figure 12.4 Typical sewer ejector setup. (*Courtesy of A. Y. McDonald Mfg. Co.*)

not have triggered the pump, unless the sump was already holding water.

When I flushed the toilet, the pump cut on. The pump should have reduced the level in the sump to where the next flush would not make the pump cut on. I flushed the toilet again, and the pump cut on immediately. This told me that there was a problem with the float.

I removed the cover from the sump, and sure enough, the float was out of position. The float cable was attached too low on the pipe in the sump. Because of its low position, the pump was forced to cut on sooner than it should.

I removed the float cable and reinstalled it. Once I had the cable tie holding the float cable in place, I reconnected the piping and replaced the sump cover. We flushed the toilet numerous times, and the problem was solved.

On a different job I was faced with the opposite problem. In this case the sewer pump wasn't cutting on soon enough. The customer knew the pump wasn't working right, so she was using the manual override to drain the sump periodically. Once she got tired of working the pump manually, she called me.

When I removed the cover from her sump, I found that the float was hitting the side of the sump and failing to rise to a level that would cut the pump on. I repositioned the cable on the float and the problem was solved.

These are just two of the common situations that plumbers run into when working with sewer ejectors. Gate valves can be closed, check valves can stick open or closed, impellers can be jammed, and all sorts of other problems can arise. Let's take some time now to look at various problems you may run into with sewer ejectors.

Vents

All sewer sumps should have vents that rise to open air, but not all do. I've seen sewer sumps in basements that were vented with only mechanical vents. Not only are these types of installations in violation of the plumbing code, they can cause real problems.

The contents of sewer sumps can produce some serious odors. Gas in the sump is another consideration. If the sump is not vented properly, gas can accumulate in it and present some potential dangers.

Mechanical vents do a good job of helping individual fixtures drain better, but they don't do much for sewer sumps. If you run across a sewer sump that is not properly vented to outside air, advise your customer of the danger and the need to vent it properly.

Gate valves

Gate valves are a part of every sewage ejector system. Normally, gate valves don't give people much trouble, but sometimes they get closed inadvertently. If a gate valve on a sewage ejector system is closed, the pump cannot empty the contents of its sump, and this is a serious problem. If you respond to a call where a sewer pump runs but doesn't seem to be pumping, check the gate valve to confirm that it is open.

Check valves

Check valves used with sewage ejectors are subject to the same problems as those described for sump pumps. Closed check valves don't allow sewer pumps to pump the contents from their sumps, and open check valves can make a pump cycle too frequently.

Floats

Floats on sewer ejectors are also subject to the same basic working principles as those found on sump pumps.

Impellers

The impellers in sewer pumps are subject to clogging and jamming. If improper objects are flushed down toilets that are connected to sewer sumps, impellers can have big problems. Moist towelettes, sanitary napkins, and similar items should never be flushed into a sump system; they can jam impellers very quickly.

Tank gaskets

Tank gaskets are an aspect of sump systems that many plumbers fail to give much credit to. These seals are responsible for keeping gas and odors confined in the sump and in the waste and vent system. If the seals deteriorate, sewer gas and odors can escape into living space. This is not only an unpleasant experience, but it can be a health hazard.

If you have a customer complaining about odors in their living space, check the gaskets on the sewer sump. You may be surprised to find that the plumber who installed the system never installed the tank-cover gasket. When the tank-cover gaskets and rubber grommets are not in good shape, odors and gas are able to escape into living space.

Basin Waste Pumps

Basin waste pumps (Figs. 12.5 and 12.6) work on the same basic principle as that used by sewer ejectors. By that, I mean that basin waste pumps are installed in sumps, are submersible, and pump water from the basin to a gravity-type drain.

Basin pumps typically have motors with $\frac{1}{3}$ horsepower, are installed in a 5-gallon sump, have an $1\frac{1}{2}$-inch inlet and a 2-inch vent. The discharge outlet for these pumps is usually of a 2-inch diameter. The electrical cables on these pumps are typically 10 feet long.

These small pumps are not short on pumping power. For example, a typical basin pump can pump 1050 gallons of water in 1 hour with a 20-foot lift, and that's a lot of water.

Basin-type pumps are used to handle water discharge from laundry trays, wet bars, water softeners, dehumidifiers, and washing machines, among other gray-water fixtures.

Basin pumps, or sink-tray pumps as they are often called, are normally installed above floor level. Sewer ejectors are installed below floor level, but these gray-water pumps can be installed in base cabinets or other above-floor levels.

Grinder Pumps

Grinder pumps (Fig. 12.7) are used for big sewage removal jobs. These pumps are powerful and can easily handle tough drainage jobs.

▼ *Sump, Sewage, Effluent and Grinder Pumps*

Effluent Pumps & Sewage Ejectors

DIMENSIONS

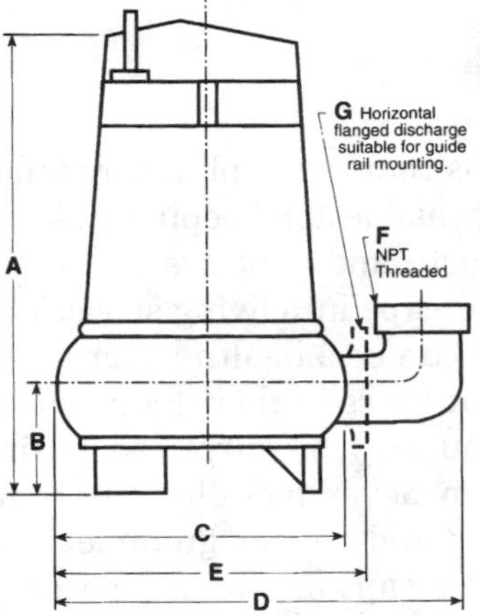

	MODEL NUMBER	A	B	C	D	E	F	G
EFFLUENT PUMPS	104011EF*	11	2 1/4	7 3/4	9	9 1/4	2	—
	105011EF*	14 1/2	2 1/4	8	11	9 1/4	2	1 1/4
	110012EF	13 3/4	2 1/4	8	11	9 1/4	2	1 1/4
	110032EF	12	2 1/4	8	11	9 1/4	2	1 1/4
	120012EF	15 3/4	2 1/4	8	11	9 1/4	2	1 1/4
	120032EF	14 1/2	2 1/4	8	11	9 1/4	2	1 1/4
STANDARD SEWAGE EJECTORS	104011SJ*	14	2 3/4	8 1/4	10	10 1/2	2	—
	105011SJ*	17	3 3/4	9	12	10 1/2	2	2
	105032SJ	14 1/2	3 3/4	9	12	10 1/2	2	2
	110012SJ	16 1/4	3 3/4	9	12	10 1/2	2	2
	110032SJ	15	3 3/4	9	12	10 1/2	2	2
	115012SJ	18 1/2	3 3/4	9	12	10 1/2	2	2
	120032SJ	17 1/4	3 3/4	9	12	10 1/2	2	2
ENGINEERED SEWAGE EJECTORS	110012EP	16 1/4	4 3/4	9 1/4	15 1/2	10 1/2	—	3
	110032EP	16	4 3/4	9 1/4	15 1/2	10 1/2	—	3
	115012EP	19 1/4	4 3/4	9 1/4	15 1/2	10 1/2	—	3
	120032EP	18 1/4	4 3/4	9 1/4	15 1/2	10 1/2	—	3
	130032EP	18 1/4	4 3/4	9 1/4	15 1/2	10 1/2	—	3

G Horizontal flanged discharge suitable for guide rail mounting.

F NPT Threaded

*U.L. Listed, C.S.A. Approved

SPECIFICATIONS

	MODEL NUMBER	RATED H.P.	SPEED R.P.M.	SOLID SIZE INCHES	OPERATING VOLTAGE	HERTZ	PHASE	DISCHARGE CONNECTION INCHES	WEIGHT	CABLE LENGTH FT.
EFFLUENT PUMPS	104011EF	4/10	3450	3/4	115 or 230	60	1	2	35	15
	105011EF	1/2	3450	3/4	115/230	60	1	2	61	15
	110012EF	1	3450	3/4	230	60	1	2	66	15
	110032EF	1	3450	3/4	230/460	60	3	2	54	15
	120012EF	2	3450	3/4	230	60	1	2	75	15
	120032EF	2	3450	3/4	230/460	60	3	2	73	15
STANDARD SEWAGE EJECTORS	104011SJ	4/10	1750	2	115 or 230	60	1	2	40	15
	105011SJ	1/2	1750	2	115/230	60	1	2	66	15
	105032SJ	1/2	1750	2	230/460	60	3	2	61	15
	110012SJ	1	1750	2	230	60	1	2	78	15
	110032SJ	1	1750	2	230/460	60	3	2	68	15
	115012SJ	1 1/2	1750	2	230	60	1	2	87	15
	120032SJ	2	1750	2	230/460	60	3	2	83	15
ENGINEERED SEWAGE EJECTORS	110012EP	1	1750	2 1/2	230	60	1	3	83	15
	110032EP	1	1750	2 1/2	230/460	60	3	3	76	15
	115012EP	1 1/2	1750	2 1/2	230	60	1	3	88	15
	120032EP	2	1750	2 1/2	230/460	60	3	3	85	15
	130032EP	3	1750	2 1/2	230/460	60	3	3	92	15

Figure 12.5 Specifications for effluent pumps and sewage ejectors. (*Courtesy of A. Y. McDonald Mfg. Co.*)

These pump systems are typically rigged with an alarm system and a total of three floats. Voltage requirements for an average grinder pump is normally 230 volts.

▼ *Sump, Sewage, Effluent and Grinder Pumps*

Sink Tray System

For maximum results where gravity drainage is not possible or practical

The "Sink Tray System" is specifically designed for residential use in pumping waste water from washing machines, laundry trays, dehumidifiers, wet bars, water softeners, etc. The system comes fully assembled and installs above the floor eliminating the need to dig a sump.

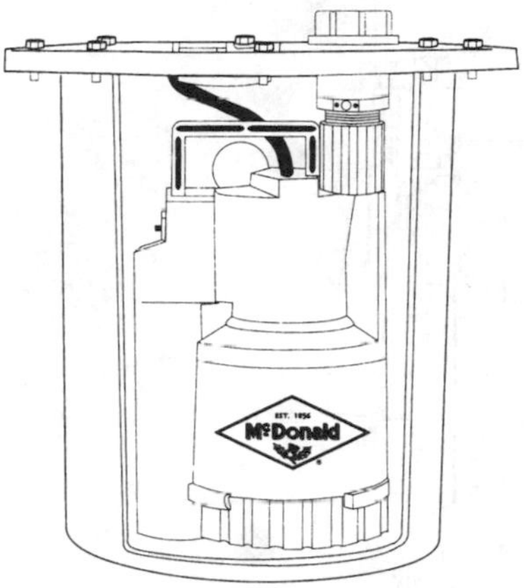

Special Features and Benefits

- Fully assembled, lightweight, self-contained and compact for easy installation.
- Installs above the floor.
- Pump includes internal float switch and check valve for trouble-free automatic operation.
- Full 1/3 HP pump for up to 1/2" solids handling.
- 10 ft. power cord with 3 prong grounded plug.
- 5 gallon corrosion resistant polyethylene tank with 1 1/2" inlet and 1 1/2" discharge compatible with all standard plumbing fittings.
- Fully vented basin for safe operation.
- Basin design ensures proper sealing for gas tight assembly.
- Quick draining action-pumps 1050 gallons of water per hour at 20 feet of lift; 2250 gallons per hour at 10 feet of lift.

Eliminates excess water from:
- Laundry trays
- Dehumidifiers
- Washing machines
- Wet bars
- Water softeners

Package Includes:
- 1/3 HP automatic totally submersible sump pump
- 5 gallon basin
- Lid
- Complete hardware package

TANK SPECIFICATIONS	
Gallons 5	Height ....15"
Inlet ....1 1/2"	Width ....15"
Vent ....2"	Weight ....16 lbs.
Discharge ....1 1/2"	

PUMP SPECIFICATIONS	
HP 1/3	Maximum fluid temp.
Voltage 115	......125° Continuous
Amp draw 2.3	......210° Intermittent
Cable length ....10 ft.	

103011SPK Performance Curve

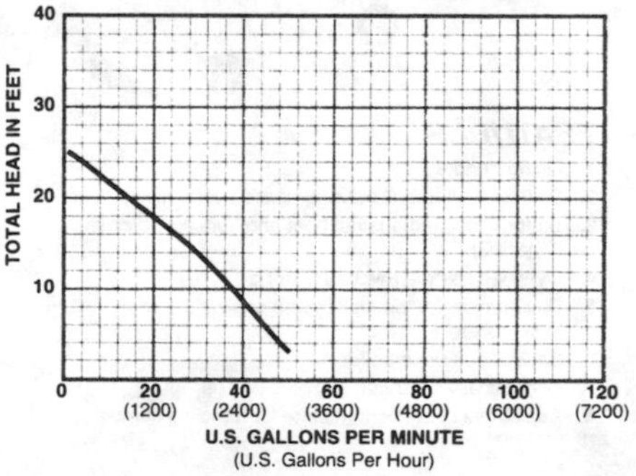

TOTAL HEAD IN FEET

U.S. GALLONS PER MINUTE
(U.S. Gallons Per Hour)

Figure 12.6 Typical basin pump setup. (*Courtesy of A. Y. McDonald Mfg. Co.*)

The discharge outlet for a typical grinder pump has a diameter of $1\frac{1}{4}$ inches, and these pumps are usually rated at 2 horsepower. In addition to the normal sump setup, grinder pumps are equipped with a control box. The control box is frequently equipped with a control

▼ *Sump, Sewage, Effluent and Grinder Pumps*

Grinder Pumps

The **A.Y. McDonald Mfg. Co. Grinder Pump** comes as a completely packaged simplex system. It is designed for residential and small industrial sewage or sump applications. The pump is recommended for homes in isolated or mountainous areas, and for dewatering of dwellings located in inland protected areas where septic tanks are not permitted. System includes pump, tank, cover, check valve, discharge piping, control box and float switches.

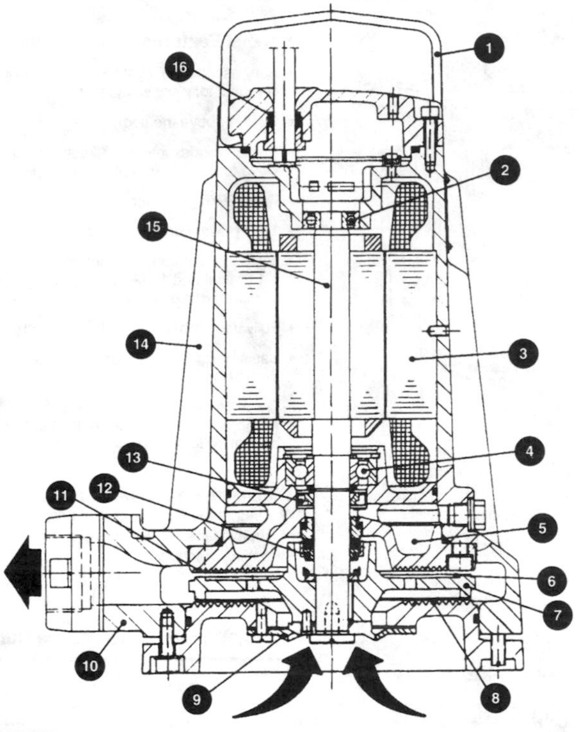

FEATURES

1. Lifting handle
2. Self-lubricating ball bearing-needs no servicing
3. Stator, insulated against heat and humidity to class F (155°C)
4. Oversized single row ball bearing
5. Oil chamber for lubrication and cooling the seal assemblies
6. Back vanes on impeller
7. Dynamically balanced impeller
8. Adjustable spiral bottom plate for handling fibrous material
9. Patented hardened rotor and stator cutter elements (Rockwell C 58-62)
10. Volute with centerline discharge suitable for mounting to guide rail bracket or discharge elbow
11. Spiral back plate
12. Mechancial lower seal enclosed in Buna N boot
13. Upper lip seal angle mounted for long life
14. Motor housing with large cooling fins
15. Rotor shaft assembly dynamically balanced
16. Watertight cable joint with strain relief

Figure 12.7 Sewage-ejector setup with alarm system. (*Courtesy of A. Y. McDonald Mfg. Co.*)

panel, hand-off automatic switch, terminal strip, and audible alarm (Fig. 12.8).

The three floats installed with a grinder pump are all set at different levels. The lowest float is the one that cuts the pump off. Floats

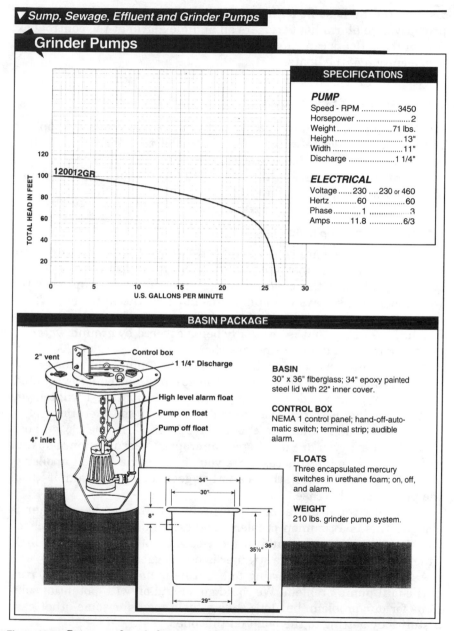

▼ *Sump, Sewage, Effluent and Grinder Pumps*

Grinder Pumps

SPECIFICATIONS

PUMP
Speed - RPM 3450
Horsepower 2
Weight ....71 lbs.
Height ....13"
Width ....11"
Discharge ....1 1/4"

ELECTRICAL
Voltage ......230 ....230 or 460
Hertz 60 60
Phase............1 ,.................3
Amps ....11.8 ....6/3

(Chart labeled 120012GR — TOTAL HEAD IN FEET vs U.S. GALLONS PER MINUTE)

BASIN PACKAGE

2" vent — Control box
— 1 1/4" Discharge
— High level alarm float
— Pump on float
— Pump off float
4" inlet

34"
30"
8"
35½" 36"
29"

BASIN
30" x 36" fiberglass; 34" epoxy painted steel lid with 22" inner cover.

CONTROL BOX
NEMA 1 control panel; hand-off-automatic switch; terminal strip; audible alarm.

FLOATS
Three encapsulated mercury switches in urethane foam; on, off, and alarm.

WEIGHT
210 lbs. grinder pump system.

Figure 12.8   Cutaway of a grinder pump. (*Courtesy of A. Y. McDonald Mfg. Co.*)

installed between the lowest and highest floats are the ones that cut the pumps on.

The float mounted nearest to the top of the sump is the alarm float. This float does not come into play unless the content level of the sump rises to the highest float, indicating that the pump is not working properly. The alarm float triggers an audible alarm at the control box. Any of these floats can cause a pump to fail, so don't forget to inspect the operation of the floats.

Troubleshooting Scenarios

Let's take a look at some troubleshooting scenarios that may apply to your work with waste-water pumps. Most of the stories that follow are based on actual case histories that I have dealt with. Some of them are fairly unique, so I think you will enjoy seeing what you may have to look forward to in the field.

A sump pump's sump is flooding

In our first case we are called to a house where a sump pump's sump is flooding. The customer has explained that she has never had this type of problem in the past, but her sump started flooding over the weekend. She also says the pump runs but that the water doesn't get pumped out as fast as its coming in. The result of this problem is a finished basement that is gradually being covered by ground water.

You arrive at the house and are shown to the basement. As you go down the stairs you can hear the sump pump running. The submersible pump is running, but the water in the sump is not being evacuated. What are you going to do? To help you in your evaluation, I will tell you that the weather is warm, so a frozen drain is not a consideration.

Are you thinking that the check valve is stuck in the closed position? If you are, you are making an appropriate assumption, but that is not the problem in this case. Are you about to check the impeller to see if it is jammed? Well, that's another good idea, but again, it is not the problem in this case.

What else could be causing the problem? Maybe the strainer is clogged; this is a common problem. You can pull the pump and check the strainer, and you should in most cases, but I'll save you a little time by telling you that the strainer is not clogged in this case.

As you are probably starting to see, this is not your common, run-of-the-mill pump problem. We could go on and on with potential solutions for our problem, but let's save a little room for some other case histories by getting to the point on this one.

In this particular case you could exhaust all of your troubleshooting skills within the home and never find the problem. The cause of all the trouble is outside.

This sump pump has one of the flexible drain hoses that are so popular. It runs underground to a small gravel-lined pit. It seems that the end of the drain pipe became somewhat of a lure to a 5-year-old, who thought it would be a grand idea to fill the pipe with rocks. And that's exactly what she did.

The corrugated ridges on the pipe worked to hold the larger rocks in place tightly. With the pipe filled with rocks, the pump could not discharge the contents of its sump. Finding this problem was not simple, but it only took a little logic and some process of elimination.

Not all pump problems are what you expect them to be. This example has shown you how you could look at every conceivable plumbing aspect of a pump problem and still never find the root cause of the problem. Not many plumbers would expect to find rocks stuffed into the end of the drain pipe. With this mystery out of the way, let's move onto the next story.

It won't quit running

What do you do with a pump when it won't quit running? This story could apply to a sump pump, a basin pump, an effluent pump, or a sewer pump. The symptoms and results for the following problem would be the same for any of the pumps. In the real job that this story is based on the pump was a sewer pump, so we will use that example in this story.

You are called to a house where the customer is complaining that his sewer pump seems to be running almost all the time. You go into the basement and observe the pump basin. The pump is running. What do you suppose the problem is?

Do you think it is a problem with the float? It could be, but in this case it is not. Are you thinking that the check valve is stuck in the open position? Well, you're getting warm, but you are not quite on target. What could it be?

The problem is with the check valve, but it is not that the check valve is stuck open. The check valve has been installed backward. This flaw in the installation is preventing the pump from removing the contents of the sump.

This situation may seem hard to believe, but I have rolled on many calls where this was the case. Unfortunately, some of the installations were done by plumbers in my employ. It is a mistake to install a check valve backward, but don't overlook a possibility you shouldn't overlook.

It runs, but is not pumping

What do you do with a pump when it runs but isn't pumping? There are a number of possible problems that will give this symptom, but if you use wise troubleshooting procedures, you can reduce wasted time in finding the cause of the problem.

The first thing to do is to disconnect the discharge piping and turn the pump on. If water is pumped under these conditions, you can direct your attention to potential problems with the piping. When water doesn't shoot up out of the pump, you can expect to find a clogged strainer or a jammed impeller. The problem will most likely be with the strainer.

If you have a pump that is not pushing water out of its discharge outlet, pull the pump and check the strainer. If the strainer is clear, concentrate on the impeller.

The alarm is going off

When the alarm is going off on a sewage pump system, most home-owners panic. When you arrive on such a job, what are you going to check first? The first thing to check is the alarm float; it may be stuck and malfunctioning.

Alarm floats are the highest floats in a sump basin, and they shouldn't send out false alarms. Unless the liquid level in the sump has risen to a point to push the alarm float up, the alarm shouldn't be sounding off. You can tell, however, if the problem is a faulty float or a problem pump by removing the cover on the sump pit.

If you look into the sump and see that the liquid level is well below the alarm float, you can assume the problem is with the float. However, when the liquid level is high, you must investigate the pump.

Pump stations for sewage removal in commercial properties should be equipped with a dual-pump system, but residences generally have only one pump. If the pump is not working properly, there are several types of problems you may have to troubleshoot. Let's look at them on a one-by-one basis.

A plugged-up pump is sure to set off the safety alarm on a pump station. Checking for this type of trouble is dirty work, but it must be done. You may get a good feel for the pump problem just by listening to it run. If the impellers are jammed, you may be able to hear a dis-crepancy in the normal running condition of the pump.

If the pump sounds like it is running properly, check the gate valve on the discharge line. Make sure it is open.

The check valve on the system could be what's causing the problem. If it is stuck in the closed position or installed backward, the pump will not be able to function properly.

One quick way to determine if the problem is in the pump or the drainage line is to disconnect the piping and observe the pump when it is running. This will tell you quickly if the pump is pumping. Be prepared to cut the pump off quickly though; otherwise you may have a large mess to clean up.

If the problem is in the pump itself, it will probably be a jammed impeller. This will require pulling the pump and doing a visual inspection. If there is a clog at the intake on the pump, you should be able to clear it with minimal trouble.

Should you establish that the problem is in the discharge piping, there are several possibilities you must consider. The gate valve is not likely to be closed, but it could be. Check valves don't usually stick in the closed position and if the system has worked properly in the past, you can rule out the possibility that the check valve is installed backward.

What does this leave you with in the way of options? If the pump is working properly and the gate and check valves are not at fault, you have some type of stoppage in the drain line or a septic system that will not accept any more discharge.

Septic systems do sometimes fill to the point where they will not accept any further discharge. If the property you are working with dumps its waste into a septic tank, remove the tank cover and inspect the liquid level. If it is above the inlet of the tank, you will have found your problem.

If the septic tank is not full, you will have to snake the drain line. The problem could be in the discharge line from the sump or in the main building drain or sewer.

Getting into the drain line to snake it will create some mess. When you remove the check valve, whatever is standing in the vertical pipe above the check valve is going to come out of the pipe, so be prepared for this. Once you have access to the pipe, you can snake it as you would any other drain.

Direct-Mount Pumps

Direct-mount pumps are the only common type of waste-water pump that we haven't discussed in this chapter. So, we will talk about them now.

Direct-mount pumps mount directly on the drains of fixtures, such as laundry tubs, bar sinks, and similar fixtures. These pumps are not as dependable or as efficient as basin-type pumps, but they are common and effective in many cases.

One of the biggest problems with direct-mount pumps is their small drain diameters. These little pumps usually have only a $\frac{3}{4}$-inch discharge outlet. If only pure water is being pumped, this is fine, but if other objects get down the drain, problems are likely to arise.

When direct-mount pumps are used on laundry tubs, they can often become jammed with foreign objects. Sand, gravel, and other small objects sometimes find their way down the drains of laundry tubs. These little particles can clog up the works in direct-mount pumps. Lint and string are also prime enemies of these small pumps. It doesn't take much to jam up the pump or the check valve used on the tiny drain lines.

If you have a direct-mount pump that is not pumping properly, inspect the check valve and the pump. You may be surprised at some of the objects you retrieve from these locations. I have found string, lint, gravel, charcoal from fish-tank filters, and other relatively common items in these locations. In the strange category, I have found plastic knives, earrings, rings, necklaces, dead goldfish, and countless other objects.

Troubleshooting Checklists

Troubleshooting checklists can help you with any kind of troubleshooting. If you can go down a checklist, there is less likelihood that you will overlook an important step in the process. You can use the following tips to help you make a troubleshooting checklist for waste-water pumps.

Pump runs but doesn't pump water

When a pump runs but doesn't pump water, there are a few things you should always look for. Let's see what they are.

- The strainer is plugged up.
- The impeller is jammed.
- The check valve is stuck in the closed position.
- The check valve is installed backward.
- The drain pipe is stopped up.
- The septic tank is full.

Pump doesn't run

When a pump isn't running at all, there are a few possibilities to consider. What are they? Well, let's see.

- A fuse is blown.
- A circuit breaker is tripped.

- The pump is not plugged in.
- The float is stuck in the off position.
- The motor is bad.
- The switch is bad.

Pump runs too often

When the pump runs too often, there are only a few things to check out; let's examine them.

- The check valve is stuck in the open position.
- The float is out of adjustment.
- There is no check valve installed.

Simple Solutions

Simple solutions are often all that are required to repair waste-water pumps. These pumps are not difficult to understand or troubleshoot, but they can send out mixed signals. If you are willing to take your time and use a pragmatic approach, finding problems with waste-water pumps will not be difficult.

As we end this chapter, we are ready to move on to the next one. In Chap. 13 we are going to delve into specialty fixtures. If you have ever wondered about fluid-suction devices, bedpan washers, water coolers, or other specialty fixtures, you are going to enjoy the next chapter.

13

Troubleshooting Specialty Fixtures

When we talk about troubleshooting specialty fixtures, we are talking about fixtures that are not common in every plumbing installation. For example, fluid-suction devices are common in dental offices, but they are not the type of plumbing device most plumbers service on a routine basis. When was the last time you had to work on the air or drain lines in a dental office?

Drinking stations are found in a lot of business and commercial situations, but they are not normally installed in homes. Residential plumbers making the transition into commercial work can feel at a loss when it comes to troubleshooting drinking stations or other commercial fixtures.

The problem with troubleshooting specialty fixtures is that they are special. Some of the specialty equipment is so special that you should have specifications from the manufacturer before working on it. This creates a problem for the plumber in the field who doesn't have a compilation of specifications from various manufacturers. It also creates a problem for authors who wish to help you learn to troubleshoot specialty fixtures.

Since there are so many different types of equipment available, I can't begin to give you step-by-step instructions for each type. I would like to, but I can't. What I can do, however, is go over the basics with you. As long as you develop a strong foundation of solid troubleshooting skills, you can overcome most problems. In some cases you will need specifications on the piece of equipment you are servicing, but your ingrained troubleshooting knowledge will carry you a long way in knowing where to look in the specifications for the answers you need.

What types of specialty fixtures and devices are we going to look at? We are going to start with a quick course in what you might expect to find in a dental office. From there, we will move on to drinking stations. After that, we will examine knee-kick devices, laboratory fittings, and more. But for now, let's concentrate on dental equipment.

Dental Equipment

I find that the most disturbing factors in working on dental equipment are the sounds that are always present in dental offices during working hours. There is something about hearing those drills running that makes it hard for me to concentrate on plumbing. Once the surrounding sounds are blocked out, I can get down to business.

You may not get squeamish at the sound of dental devices doing their work, but do you know what to look for when problems arise with dental equipment? I would venture a guess that only a very small percentage of plumbers have ever worked on dental equipment.

During my career, I have installed plumbing in new buildings for dental care, and I've done service and repair work on a fair amount of dental equipment. It is not the kind of work I do every day, but I find myself around it often enough to keep up with most of the changes in the industry.

The last new installation I did for a dental office involved a bottle-type fluid-suction system. At that time, bottle receptors were common practice. I'm finding now that fewer dental facilities are using the bottle systems in my area. Instead, they are equipped with suction equipment that dumps the waste into a drain connected to the sanitary sewer.

The fact that human fluids are being discharged into the sanitary sewer bothers me a little. With all the new fears of the diseases that plumbers can be subjected to, I don't like the idea of cutting open a drain pipe, thinking that is carries only normal sewage and waste, to discover the types of fluids put into the pipe from a suction system.

If I were in charge of making the plumbing rules and regulations, I would insist that there be some way for plumbers to positively identify pipes that are carrying waste other than normal sewage and graywater waste. But, I don't make the rules, so be advised. When you open a drain in a building that houses a medical office, you may find more than you bargained for in it.

When you are called in to service equipment in dental offices, you may be working with air, drain, or water lines. You can troubleshoot the primary water and drain pipes in the same way you would any other water distribution or drainage, waste and vent system. The supply piping to, and the drainage piping from, the dental equipment is a little different. And then there are the air lines. To put this into perspective, let's look at a couple of standard setups that you may be called upon to work with.

Air piping

The air piping from a compressor to the workstation at a dental chair is not complicated. The system begins with an air compressor. From the compressor, a ½-inch air line is run. It is usually piped with type L copper tubing. In most cases, the supply pipe from the compressor is run to another piece of equipment that removes moisture from the air.

When the air piping leaves the second piece of equipment, it is run to the workstation. Typically, there will be a high-pressure cutoff switch and a quick-opening valve where the supply piping meets the work area. It is also common for the pipe size to be enlarged to ¾ inch when it arrives at its final destination.

Very few plumbers, if any, take responsibility for troubleshooting the special equipment in these setups. Normally, plumbers are only looked to for assistance in piping problems. This basically limits the knowledge needed from the plumber to the repair of leaks.

If you work with this type of equipment enough, you will find that the air compressor is likely to be equipped with its own drain. This drain enables the tank to evacuate the buildup of condensation at the end of each pumping cycle.

Evacuation systems

Evacuation systems in dental offices can come in many forms, but we will discuss one of the more current and more common ones. In this evacuation system you will run into a few things you are probably not accustomed to seeing.

The system begins at the dental chair, with the suction hose. The hose normally has a diameter of 1 inch. As the hose runs from the chair toward the suction equipment, it passes through an in-line filter. It then enters the equipment.

On the opposite side of the equipment you will find a ½-inch cold-water supply. This supply will have a cutoff valve, and the ½-inch tubing will be reduced to ⅜-inch hose. The ⅜-inch hose is what connects to the equipment.

Below the water inlet there will be a discharge outlet. A 1-inch hose is run from the outlet, through a muffler, to a 1½-inch drain. The connection at the drain is often made with a Durham P-trap and a solid connection, not an open air-gap.

Keep in mind that the examples I have just given you are not the only configurations you are likely to find in dental offices. The age of the equipment will have a bearing on what you run into and so will the brands and models of the equipment.

Some vacuum systems require the same 1-inch piping that I described above, except that their requirements call for the piping to be done with schedule 40 PVC.

A master control with an in-line filter may also be installed. These devices allow the dentists to shut down the water supply to their equipment by throwing one switch. One of the purposes of this unit is to prevent leaks at the work area. The other is to filter water being supplied to the dental tools. These devices come in both ½- and ¾-inch sizes.

As intimidating as working in a dental office may be, the troubleshooting of the plumbing is not that different from any other type of troubleshooting you do. As long as you use a structured system to locate the cause of problems, you shouldn't have any unusual trouble when working with dental equipment. However, I strongly recommend that you obtain and refer to manufacturer's recommendations and specifications when working on any specialty equipment.

Drinking Stations

Drinking stations are, in my opinion, something of a specialty fixture; that is why we are going to discuss them in this section of the book. Since drinking stations are not standard equipment in all plumbing systems, there are many plumbers who have only a limited knowledge of how to troubleshoot them. Even though I'm calling drinking stations specialty fixtures, they are abundant enough to deserve some attention.

People often call drinking stations drinking fountains. They also refer to bubblers as drinking fountains. This can get confusing, since there are coolers, fountains, and bubblers. It can be important to know which type of drinking station you are being asked to work on. We are going to start our troubleshooting tour with drinking fountains. To aid you in your troubleshooting of water coolers and fountains, I'm including numerous drawings that illustrate the different types of equipment you may have to work with (Figs. 13.1 through Fig. 13.19)

Standard drinking fountains

The drain for an average drinking fountain has a 1¼-inch diameter. Sometimes these drains are trapped directly below the fixture, and sometimes the drain offsets into a wall before it is trapped. The drain system for a water fountain can be treated, in terms of troubleshooting, in a way similar to that used on lavatories.

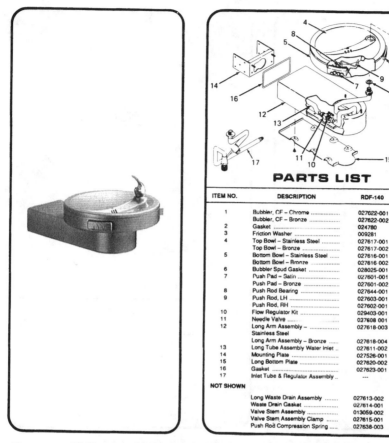

PARTS LIST

ITEM NO.	DESCRIPTION	RDF-140	FLF140R
1	Bubbler, CF – Chrome	027622-001	030951-001
	Bubbler, CF – Bronze	027622-002	030951-002
2	Gasket	024780	030017-001
3	Friction Washer	009281	009281
4	Top Bowl – Stainless Steel	027617-001	027617-001
	Top Bowl – Bronze	027617-002	027617-002
5	Bottom Bowl – Stainless Steel	027616-001	027616-001
	Bottom Bowl – Bronze	027616-002	027616-002
6	Bubbler Spud Gasket	028025-001	028025-001
7	Push Pad – Satin	027601-001	027601-001
	Push Pad – Bronze	027601-002	027601-002
8	Push Rod Bearing	027644-001	027644-001
9	Push Rod, LH	027603-001	027603-001
	Push Rod, RH	027602-001	027602-001
10	Flow Regulator Kit	029403-001	---
11	Needle Valve	027608-001	---
12	Long Arm Assembly – Stainless Steel	027618-003	027618-003
	Long Arm Assembly – Bronze	027618-004	027618-004
13	Long Tube Assembly Water Inlet	027611-002	---
14	Mounting Plate	027526-001	027526-001
15	Long Bottom Plate	027620-002	027620-002
16	Gasket	027623-001	027623-001
17	Inlet Tube & Regulator Assembly	---	029519-004
NOT SHOWN			
	Long Waste Drain Assembly	027613-002	027613-002
	Waste Drain Gasket	027614-001	027614-001
	Valve Stem Assembly	013059-002	013059-002
	Valve Stem Assembly Clamp	027615-001	027615-001
	Push Rod Compression Spring	027638-003	027638-003

Figure 13.1 Wall-mount drinking fountain. (*Courtesy of EBCO Mfg. Co.*)

The water supply for a typical drinking fountain will come out of a wall and connect to the unit. A ½-inch supply is usually run to the cutoff valve, where the supply is reduced to ⅜-inch tubing.

In terms of possible complications with drinking fountains, there are many considerations. For example, mineral deposits can build up and restrict the flow of water. The flow-regulator kit in the unit can require attention to compensate for an unsatisfactory flow of water. Then the inlet tube and regulator assembly can give you trouble. Any of these options could cause a drinking fountain to act up.

Leaks are another consideration. You might have a leak where the bubbler is attached to the interior water supply. There is a bubbler spud gasket at this location you can check. The needle valve that is inside the unit could develop a leak, and, of course, a leak could come from the connections at the inlet tube and regulator assembly.

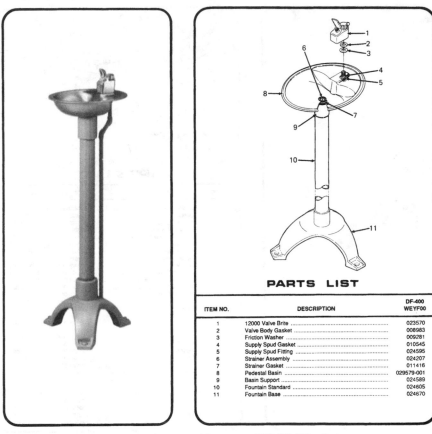

PARTS LIST

ITEM NO.	DESCRIPTION	DF-400 WEYF00
1	12000 Valve Brite	023570
2	Valve Body Gasket	008983
3	Friction Washer	009281
4	Supply Spud Gasket	010545
5	Supply Spud Fitting	024595
6	Strainer Assembly	024207
7	Strainer Gasket	011416
8	Pedestal Basin	029579-001
9	Basin Support	024589
10	Fountain Standard	024605
11	Fountain Base	024670

Figure 13.2   Pedestal drinking fountain. (*Courtesy of EBCO Mfg. Co.*)

None of the problems we have just discussed are hard to understand or to correct. As long as you know where to look, and what to look for, you can remain in control during the troubleshooting of a drinking fountain. I have provided you with some detailed drawings that show the components of a standard drinking fountain. These drawings, like others throughout this book, can be more helpful than a thousand words on the subject.

Pedestal fountains

Pedestal fountains are easy to troubleshoot in that there are not many concealed parts. The water supply runs exposed to the bubbler valve, and the only concealed parts in the water system are the gasket for the valve body and the gasket for the supply spud.

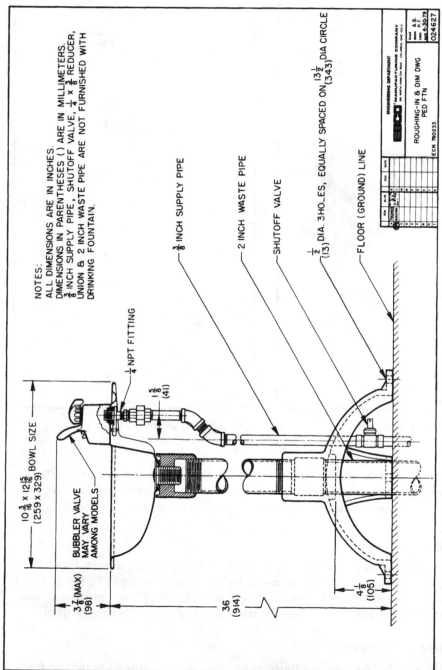

Figure 13.3 Detail of pedestal drinking fountain. (*Courtesy of EBCO Mfg. Co.*)

NOTES:
ALL DIMENSIONS ARE IN INCHES.
DIMENSIONS IN PARENTHESES ( ) ARE IN MILLIMETERS.
$\frac{3}{8}$ INCH SUPPLY PIPE, SHUTOFF VALVE, $\frac{1}{4}$ x $\frac{3}{8}$ REDUCER,
UNION & 2 INCH WASTE PIPE ARE NOT FURNISHED WITH
DRINKING FOUNTAIN.

$\frac{3}{8}$ INCH SUPPLY PIPE

2 INCH WASTE PIPE

SHUTOFF VALVE

$\frac{1}{2}$ (13) DIA. 3 HOLES, EQUALLY SPACED ON 13$\frac{1}{2}$ (343) DIA CIRCLE

FLOOR (GROUND) LINE

$\frac{1}{4}$ NPT FITTING

1$\frac{5}{8}$ (41)

10$\frac{3}{16}$ x 12$\frac{15}{16}$ (259 x 329) BOWL SIZE

BUBBLER VALVE
MAY VARY
AMONG MODELS

3$\frac{7}{8}$ (MAX) (98)

36 (914)

4$\frac{1}{8}$ (105)

247

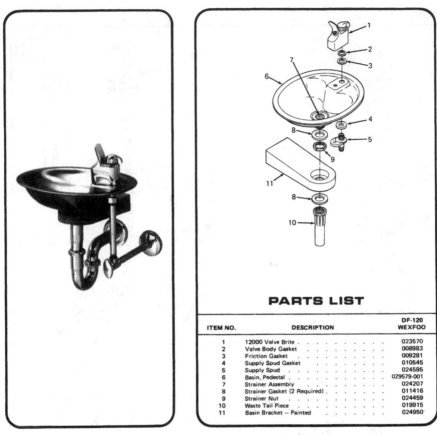

PARTS LIST

ITEM NO.	DESCRIPTION	DF-120 WEXFOO
1	12000 Valve Brite	023570
2	Valve Body Gasket	008983
3	Friction Gasket	009281
4	Supply Spud Gasket	010545
5	Supply Spud	024595
6	Basin, Pedestal	029579-001
7	Strainer Assembly	024207
8	Strainer Gasket (2 Required)	011416
9	Strainer Nut	024459
10	Waste Tail Piece	019915
11	Basin Bracket — Painted	024950

Figure 13.4 Exposed-trap drinking fountain. (*Courtesy of EBCO Mfg. Co.*)

The drainage for these types of fountains is concealed in the pedestal and trapped below the floor level. The drainage for this type of unit is designed to be connected to a 2-inch drain.

Simple bubblers

Simple bubblers are easy to work with. They hang on a wall and have an exposed trap and an exposed water supply. Aside from the bubbler valve, the only other parts to be considered, other than routine plumbing fittings, are the gaskets for the valve body and the supply spud.

True water coolers

True water coolers are more complicated than simple drinking fountains. The fact that these units cool drinking water, rather than sim-

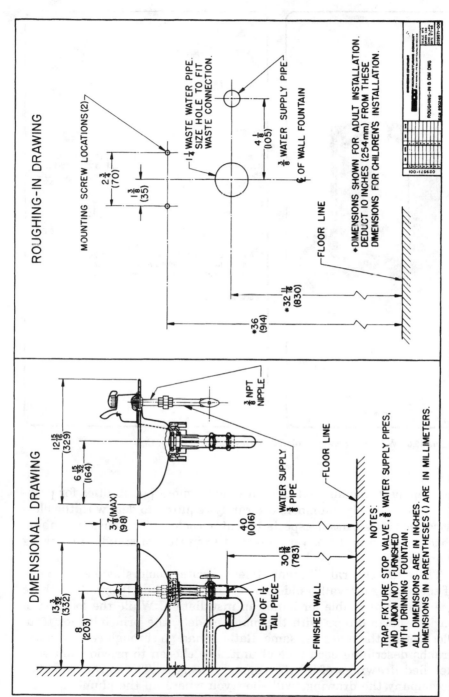

Figure 13.5 Detail of exposed-trap drinking fountain. (*Courtesy of EBCO Mfg. Co.*)

SPECIFICATIONS

Voltage	115 VAC ±10%/1PH/60 Hertz
Size (Approx)	26-1/2" H., 17-1/2" W., 18-1/2" D.
Shipping Weight (Approx)	67 LBS.
Cold Water Capacity (GPH)*	8
Compressor (HP)	1/5
Compressor (Amps) (Full Load)	6
Refrigerant	R-12
Refrigerant Charge	5.0 OZ.

* A.R.I. Rating: Room Temperature 90°F; Supply Water
Temperature 80°F; Drinking Water Temperature 50°F.
Based on standard test procedures.

Specifications subject to change without notice.

Figure 13.6 Wall-hung water cooler. (*Courtesy of EBCO Mfg. Co.*)

ply supply it, means that there are many more possibilities for problems. Fortunately, plumbers are rarely required to deal with the electrical or refrigeration aspects of water coolers. The basic plumbing
with these units is not very different from that of standard drinking
fountains.

There are several different styles of water coolers available. Some
of them hang on walls, and some of them stand on floors. And, there
are models available for handicap installation. While the location of
various parts differs with the many models, the principles for troubleshooting them are the same. Rather than go through several paragraphs describing each type of unit, I've chosen to provide you with
detailed drawings of all the common designs. The parts lists that
accompany the drawings will show you where all the plumbing components are located.

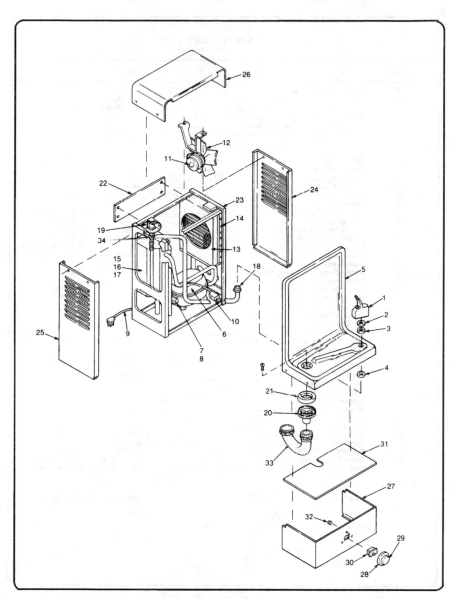

Figure 13.7   Water cooler. (*Courtesy of EBCO Mfg. Co.*)

PARTS LIST

ITEM NO.	DESCRIPTION	PLF8SKPE	PLF8SKTP
1	12000A Bubbler Valve	030774-004	030774-004
2	Valve Body Gasket	028706-006	028706-006
3	Friction Washer	028706-013	028706-013
4	Bubbler Spud Gasket	028706-001	028706-001
5	Simulated Recessed Top – Stainless Steel	028849-001	028849-001
6	Compressor Assembly (Includes Items 7 and 8)	027256-003	027256-003
7	Relay	017978	017978
8	Overload	016942-005	016942-005
9	Service Cord	026540-004	026540-004
10	Cold Control	027040-007	027040-007
11	Fan Motor	027354-009	027354-009
12	Fan Blade	022977	022977
13	Shroud	029125-001	029125-001
14	Condenser	029111-001	029111-001
15	Insulated Cooling Tank Assembly	029108-003	029108-003
16	Cooling Tank Insulation LH	028661-001	028661-001
17	Cooling Tank Insulation RH	028660-001	028660-001
18	Bubbler Supply Spud	028701-002	028701-002
19	Solenoid Valve	031011-001	031011-001
20	Waste Assembly	023046-001	023046-001
21	Waste Gasket	018050	018050
22	Hanger Bracket and Barrier Assembly	025662-003	025662-003
23	Frame Assembly	022967	022967
24	Right Side Panel – Sandstone	022979-002	022979-002
	Right Side Panel – Stainless Steel	023036	023033
	Right Side Panel – Painted	023038	023038
25	Left Side Panel – Sandstone	022980-002	022980-002
	Left Side Panel – Stainless Steel	023035	023035
	Left Side Panel – Painted	023037	023037
26	Cap – Sandstone	031068-005	031068-005
	Cap – Stainless Steel	031068-003	031068-003
	Cap – Painted	031068-002	031068-002
27	Wrapper – Sandstone	031069-204	---
	Wrapper – Stainless Steel	031069-202	---
	Wrapper – Painted	031069-201	---
	Wrapper – SAN TP	---	031069-304
	Wrapper – STN TP	---	031069-302
	Wrapper – Painted TP	---	031069-301
28	Bezel	024816	024816
29	Push Button	024817	024817
30	Switch	024764	024764
31	Support Plate	023391	023391
32	Self Threading Hex Nut	020203	020203
33	Waste Trap Assembly	027542-001	027542-001
34	Male Fitting w/Gasket – GHT X 1/4 NPT (Solenoid Valve Inlet)	031033-001	031033-001
Not Shown			
	Male Fitting Gasket	028706-024	028706-024

Figure 13.8   Parts detail of water cooler in Fig. 13.7. (*Courtesy of EBCO Mfg. Co.*)

SPECIFICATIONS

Voltage	115 VAC ±10%/1PH/60 Hertz
Size	25″ H., 18″ W., 19″ D.
Shipping Weight (Approx)	60 LBS.
Cold Water Capacity	7.8 GPH
Compressor	1/5 HP
Refrigerant	R-500
Refrigerant Charge	4 OZ.
Compressor (Full Load)	6 Amps

Specifications subject to change without notice.

Figure 13.9   Wall-hung water cooler. (*Courtesy of EBCO Mfg. Co.*)

Specialty Valves and Fittings

Next we are going to talk about specialty valves and fittings. These devices are common in some plumbing systems, such as hospitals, but rare in most. While you may never have a need to work with the equipment we are about to discuss, the background knowledge you gain from the following paragraphs may prove helpful.

Knee-action valves

Knee-action valves (Figs. 13.20 through 13.23) are most commonly found in hospitals. These specialty fittings allow the operation of fixtures without the use of hands. This allows medical personnel to avoid contamination from faucet handles, and it also leaves both hands

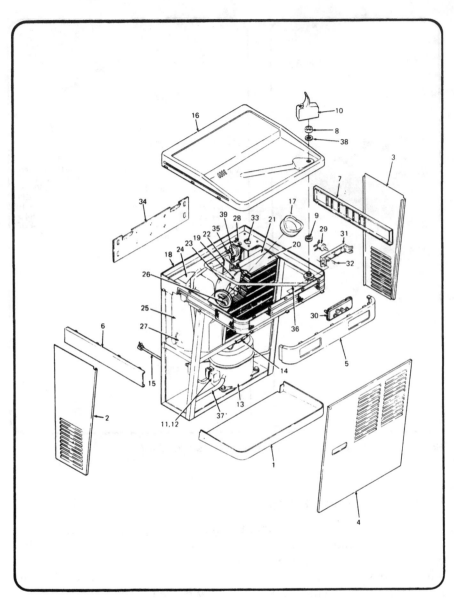

Figure 13.10 Water cooler. (*Courtesy of EBCO Mfg. Co.*)

PARTS LIST

ITEM NO.	DESCRIPTION	PLF8WMD-002	PLF8WMQ-002 PLF8WMQA-003
1	Front Housing – Sandstone	026780-002	026780-002
	Front Housing – Stainless Steel	026780-004	026780-004
	Front Housing – Tan	026780-005	026780-005
2	Left Side Panel – Sandstone	026798-003	026798-003
	Left Side Panel – Stainless Steel	026798-007	026798-007
	Left Side Panel – Tan	026798-009	026798-009
3	Right Side Panel – Sandstone	026798-004	026798-004
	Right Side Panel – Stainless Steel	026798-008	026798-008
	Right Side Panel – Tan	026798-010	026798-010
4	Front Panel – Sandstone	026802-002	026802-002
	Front Panel – Stainless Steel	026802-004	026802-004
	Front Panel – Tan	026802-005	026802-005
5	Front Bezel	026869-001 OPEN	026869-001 OPEN
6	Left Side Bezel	026886-003 CLOSED	026886-001 OPEN
7	Right Side Bezel	026886-004 CLOSED	026886-002 OPEN
8	Valve Body Gasket	028706-006	028706-006
9	Washer (Under Top)	028706-001	028706-001
10	Bubbler Valve Assembly	030774-004	030774-004
11	Overload	017977	017977
12	Relay	017978	017978
13	Compressor	027256-002	027256-002
14	Cold Control	027040-007	027040-007
15	Service Cord	026540-004	026540-004
16	Top	028835-001	028835-001
17	Waste Gasket	027205	027205
18	Frame Assembly	026976-003	026976-003
19	Fan Blade	024033	024033
20	Condenser	026967	026967
21	Fan Shroud	026968	026968
22	Fan Motor Bracket	026969	026969
23	Fan Motor Assembly	027354-009	027354-009
24	Cooling Tank Insulation (RH)	028660-001	028660-001
25	Cooling Tank Insulation (LH)	028661-001	028661-001
26	Waste Assembly	026981-001	026981-001
27	Cooling Tank Assembly (Insulated)	031045-001	031045-001
28	Solenoid Valve	031011-001	031011-001
29	Switch	028591-001	028591-001
30	Switch Pad	028587-001	028587-001
31	Switch Bracket	028588-001	028588-001
32	Compression Spring	027638-005	027638-005
33	Timer (QA models)	---	030973-001
34	Hanger Bracket	016698-002	016698-002
35	Connector Fitting	023834-002	023834-002
36	Tube Assembly Sol to Blb	031035-001	031035-001
37	Heat Exchanger	028785-001	028785-001
38	Friction Washer	028706-013	028706-013
39	Male Fitting w/Gasket –GHT X 1/4 NPT	031033-001	031033-001

NOT SHOWN

	Male Fitting Gasket	028706-024	028706-024

Figure 13.11 Parts detail of water cooler in Fig. 13.10. (*Courtesy of EBCO Mfg. Co.*)

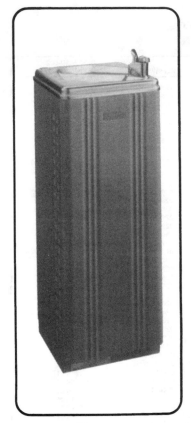

SPECIFICATIONS

7P MODELS

Voltage	115 VAC ±10%/1PH/60 Hertz
Size (Approx)	40" H., 15" W., 15" D.
Shipping Weight (Approx)	73 LBS.
Cold Water Capacity	7 GPH
Compressor	1/8 HP
Compressor Amps (Full Load)	2.5
Refrigerant	R-500
Refrigerant Charge	4-3/8 OZ.

13P AND 13PL MODELS

Voltage	115 VAC ±10%/1PH/60 Hertz
Size (Approx)	40" H., 15" W., 15" D.
Shipping Weight (Approx)	80 LBS.
Cold Water Capacity	13 GPH
Compressor	1/5 HP
Compressor Amps (Full Load)	5.0
Refrigerant	R-12
Refrigerant Charge	5 OZ.

*A.R.I. Rating: Room Temperature 90°F; Supply Water Temperature 80°F; Drinking Water Temperature 50°F.

Specifications subject to change without notice.

Figure 13.12   Floor-mount water cooler. (*Courtesy of EBCO Mfg. Co.*)

free. Knee-action valves can also be used to accommodate handicapped people.

The activator on these units can be made to mount on either a wall, the crossbar of a chair carrier, or the door of a laboratory cabinet.

The cutoff valves for knee-action valves are built into the unit. Nipples are used to connect the valves to water supplies in walls. When pressure is applied on the knee bracket, it causes the valve to open the supply channels, bringing water to the device in use. Once pressure is removed from the knee bracket, the valve closes and stops the flow of water. The action is similar to push-button faucets.

Repair kits and parts are available for the various makes and models. If you have a fixture that is not producing water in the desired manner, you can cut off the stop valves on the knee-action valve and troubleshoot the mechanisms. Once the defective part is identified, it can be replaced easily.

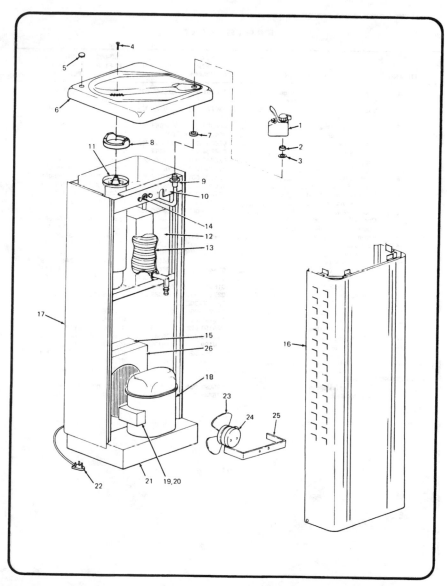

Figure 13.13 Detail of water cooler in Fig. 13.12. (*Courtesy of EBCO Mfg. Co.*)

PARTS LIST

ITEM NO.	DESCRIPTION	7P-003 PLF7P PLF7P-001	7P-001D PLF7P-D PLF7P-D001	13P-003 PLF13P PLF13P-001	13PL-003 PLF13PL PLF13PL-001
1	Bubbler Valve 12000 – Used on all models except PLF-001 Series	026551-002	026551-002	026551-002	026551-002
1	Bubbler Valve 12000A – Used on PLF-001 Series only	030774-002	030774-002	030774-002	030774-002
2	Valve Body Gasket	008983	008983	008983	008983
3	Friction Washer	009281	009281	009281	009281
4	Drain Screw	026655-002	026655-002	026655-002	026655-002
5	Hole Cover	016994	016994	016994	016994
6	Top	028845-001	028845-001	028845-001	028845-001
7	Supply Spud Gasket	028706-001	028706-001	028706-001	028706-001
8	Waste Gasket	027389-001	027389-001	027389-001	027389-001
9	Bubbler Supply Spud	028701-002	028701-002	028701-002	028701-002
10	Upper Cabinet Bracket	027407-001	027407-001	027407-001	027407-001
11	Precooler Assembly	027412-001	027412-001	027412-001	027412-001
12	Cooling Tank Insulation – 2 Rqd	028230-002	028230-002	---	---
	Cooling Tank Insulation – R.H.	---	---	028660-001	028660-001
	Cooling Tank Insulation – L.H.	---	---	028661-001	028661-001
13	Insulated Cooling Tank	029673-001	029673-001	029096-001	029096-002
14	Cold Control	027040-007	027040-007	027040-007	027040-007
15	Condenser	018252	018252	018252	018252
16	Front Panel – Tan	024442-005	024442-005	024442-005	024442-009
	Front Panel – Stainless Steel	024442-003	024442-003	024442-003	024442-011
17	Cabinet – Tan	015570-002	015570-002	015570-002	015569-002
	Cabinet – Stainless Steel	015570-005	015570-005	015570-005	015569-005
18	Compressor (Includes Items 19 and 20)	027245-001	026868-003	027256-002	027256-002
19	Overload	027268-002	---	017977	017977
20	Relay	027267-003	026867	017978	017978
21	Base – Tan	028199-001	028199-001	028199-001	028199-001
	Base – Stainless Steel	028199-002	028199-002	028199-002	028199-002
22	Service Cord	026540-003	026540-003	026540-005	026540-005
23	Fan Blade	019199	019199	022977	022977
24	Fan Motor	027354-020	027354-020	027354-008	027354-008
25	Fan Bracket	017530	017530	017530	017530
26	Fan Shroud	017821	017821	029656-001	029656-001
ACCESSORY PARTS					
	Glass Filler Installation Kit	029098-001	029098-001	029098-001	029098-001
	Foot Pedal Kit	026645-014	026645-014	026645-014	026645-013

Figure 13.14   Parts detail of water cooler in Fig. 13.12. (*Courtesy of EBCO Mfg. Co.*)

SPECIFICATIONS

Models CP3M-001D, CP3M-P, CP3M-Z,
PLF3CM, PLF3CM-D, & PLF3CM-P

Voltage	115 VAC ±10%/1PH/60 Hertz
Size (Approx)	16" H., 17" W., 13" D.
Shipping Weight (Approx)	49 LBS.
Cold Water Capacity	3 GPH
Compressor - All Models	1/8 HP
Compressor (Full Load) 115 Volt	2.5 Amps
Refrigerant	R-12
Refrigerant Charge - All Models	3-1/8 OZ.

Models CP3M-D5/6 & PLF3CMY-D

Voltage	220/240 Volt 50/60 Hertz
Size (Approx)	16" H., 17" W., 13" D.
Shipping Weight (Approx)	49 LBS.
Cold Water Capacity	3 GPH
Compressor - All Models	1/8 HP
Compressor (Full Load) 220 Volt	1.2 Amps
Refrigerant	R-12
Refrigerant Charge - All Models	3-1/8 OZ.

*A.R.I. Rating: Room Temperature 90°F; Supply Water Temperature 80°F; Drinking Water Temperature 50°F.

Specifications subject to change without notice.

Figure 13.15 Wall-hung water cooler. (*Courtesy of EBCO Mfg. Co.*)

Pedal valves

Pedal valves (Figs. 13.24 through 13.27) are used for purposes similar to those of knee-action valves. The installation of pedal valves is usually made at floor level, but wall-hung models are available. These valves can be equipped to work with a single water supply or with a dual water supply.

Pedal valves that are designed to mix hot and cold water have two inlet openings and one outlet opening. When the pedals are depressed, they cause a valve in the body to open, providing water to a fixture. These specialty fittings are easy to disassemble and troubleshoot.

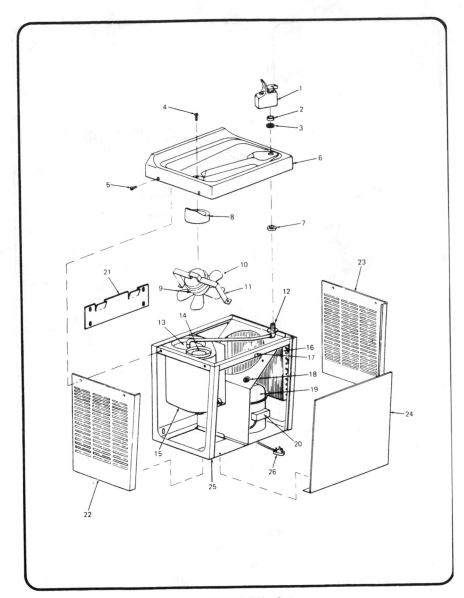

Figure 13.16   Water cooler. (*Courtesy of EBCO Mfg. Co.*)

PARTS LIST

ITEM NO.	DESCRIPTION	CP3M PLF3CM SERIES	CP3M-D5/6 PLF3CMY-D SERIES
1	12000 Bubbler Valve	026551-002	026551-002
	12000A Bubbler Valve (PLF-001 Series)	030774-002	030774-002
2	Valve Body Gasket	008983	008983
3	Friction Washer	009281	009281
4	Drain Screw	026655-002	026655-002
5	Top Screw	026630-005	026630-005
6	Top	028846-003	028846-003
7	Gasket (Under Top)	028706-001	028706-001
8	Waste Gasket	018050	018050
9	Fan Motor Assembly	027354-021	027354-006
10	Fan Blade	020243	020243
11	Fan Motor Bracket	026969	026969
12	Bubbler Supply Spud	029177-001	029177-001
13	Cooling Tank Assembly	029437-002	029437-002
14	Waste Assembly	019988	019988
15	Cooling Tank Insulation L/H	019990L	019990L
	Cooling Tank Insulation R/H	019990R	019990R
16	Condenser	028072-001	028072-001
17	Panel Clip	018156-001	018156-001
18	Cold Control	027040-007	027040-005
19	Compressor (Includes Item 20)	See Below	026883-003
20	Relay	See Below	026309
21	Hanger Bracket & Barrier Assembly	025662-002	025662-002
22	Left Side Panel – Tan	026935-003	026935-003
	Left Side Panel – Stainless Steel	026935-027	026935-027
23	Right Side Panel – Tan	026935-004	026935-004
	Right Side Panel – Stainless Steel	026935-028	026935-020
24	Front Panel – Tan	026925-001	026925-001
	Front Panel – Stainless Steel	026925-014	026925-014
25	Frame Assembly	029436-001	029436-001
26	Service Cord	026540-013	028625-002

COMPRESSORS

DANFOSS CP3M-D AND PLF3CM-D

19	*Compressor	026868-003
20	Relay	026867
	Overload	Internal

PANASONIC CP3M-P AND PLF3CM-P

19	*Compressor	028570-003
20	Relay	027709-002
	Overload	Internal

TECUMSEH CP3M-Z AND PLF3CM

19	*Compressor	027245-001
20	Relay	027267-003
	Overload	027268-002

*NOTE: Compressors include relay and overload.

Figure 13.17   Parts detail of water cooler in Fig. 13.16. (*Courtesy of EBCO Mfg. Co.*)

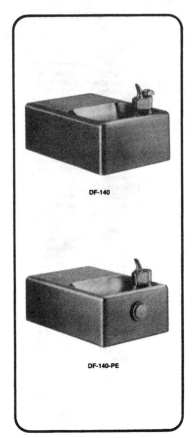

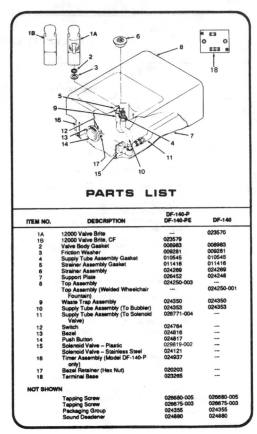

PARTS LIST

ITEM NO.	DESCRIPTION	DF-140-P DF-140-PE	DF-140
1A	12000 Valve Brite	---	023570
1B	12000 Valve Brite, CF	023579	
2	Valve Body Gasket	008983	008983
3	Friction Washer	009281	009281
4	Supply Tube Assembly Gasket	010545	010545
5	Strainer Assembly Gasket	011416	011416
6	Strainer Assembly	024269	024269
7	Support Plate	026452	024248
8	Top Assembly	024250-003	---
	Top Assembly (Welded Wheelchair Fountain)	---	024250-001
9	Waste Trap Assembly	024350	024350
10	Supply Tube Assembly (To Bubbler)	024353	024353
11	Supply Tube Assembly (To Solenoid Valve)	026771-004	---
12	Switch	024764	---
13	Bezel	024816	---
14	Push Button	024817	---
15	Solenoid Valve – Plastic	029819-002	---
	Solenoid Valve – Stainless Steel	024121	---
16	Timer Assembly (Model DF-140-P only)	024937	---
17	Bezel Retainer (Hex Nut)	020203	---
18	Terminal Base	023265	---
NOT SHOWN			
	Tapping Screw	026680-005	026680-005
	Tapping Screw	026675-003	026675-003
	Packaging Group	024355	024355
	Sound Deadener	024880	024880

Figure 13.18   Wheelchair drinking fountain. (*Courtesy of EBCO Mfg. Co.*)

Bedpan flushers

Bedpan flushers (Figs. 13.28 and 13.29) are available in numerous configurations. For example, one type of flusher is a diverter flushing fitting. It has a vertical swing spout and a removable spray outlet. This type of fitting is available with or without an antiseptic injector unit.

Another type of bedpan flusher resembles a wall-mounted laundrytub faucet. It has a vacuum breaker on top of its spout and a spray assembly.

As with most specialty fixtures, these faucets can be treated with the same troubleshooting skills given for other faucets and fixtures throughout this book.

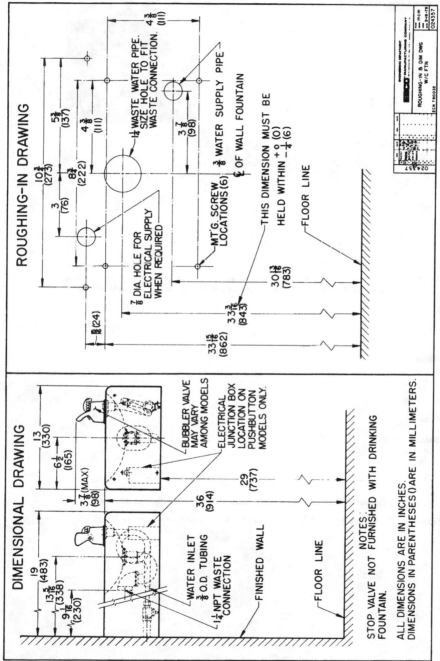

Figure 13.19 Detail of drinking fountain in Fig. 13.18. (*Courtesy of EBCO Mfg. Co.*)

263

NORWICH
Vitreous China Lavatory

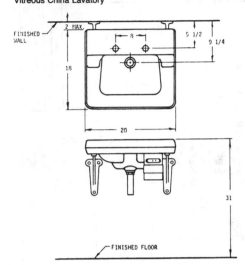

Norwich 1H-163 on Enameled Brackets

Lavatory: 1H-163 *Norwich* vitreous china lavatory with rectangular basin, splash lip, front overflow and two soap depressions. Punched for goose-neck spout, soap valve and fixture-mounted knee-action mixing valve.

Support:

Trim:

Waste:

Trap:

Supplies:

Color:

Soap Valve:

Exposed metal trim is polished chromium plated, unless otherwise described.

Figure 13.20   Lavatory with knee-action valve. (*Courtesy of Crane Plumbing.*)

NORWICH LAVATORY
VITREOUS CHINA
EXPOSED ENAMELED IRON BRACKETS
KNEE ACTION VALVE AND GOOSENECK SPOUT
PUSH BUTTON SOAP VALVE

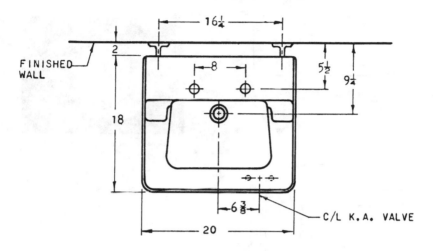

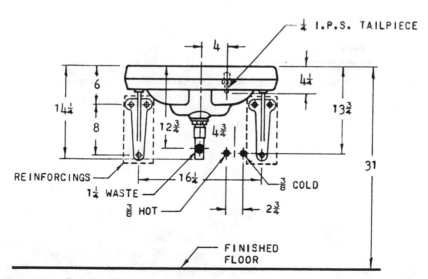

Figure 13.21 Detail of lavatory with knee-action valve in Fig. 13.20. (*Courtesy of Crane Plumbing.*)

NEU-MEDIC™
Vitreous China Surgeon's Scrub-Up Sink

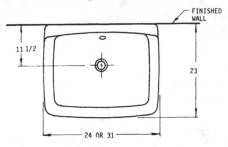

Neu-Medic 5H-256 Surgeon's Scrub-up Sink
Vitreous China

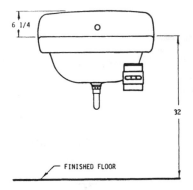

Sink:	5H-256 *Neu-Medic* vitreous china surgeon's sink with 6¼" high back and return ends.

Supports:

Trim:

Supplies:

Waste:

Trap:

Exposed metal trim is polished chromium plated, unless otherwise described.

Figure 13.22 Surgeon's scrub sink with knee-action valve. (*Courtesy of Crane Plumbing.*)

NEU-MEDIC™ WASH-UP SINK
VITREOUS CHINA
KNEE ACTION VALVE AND GOOSENECK SPOUT
OPEN STRAINER
CAST OR BENT TUBE "P" TRAP

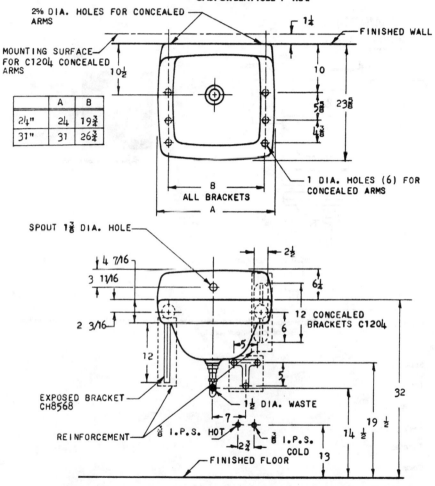

NOTE: ROUGHING-IN DIMENSIONS FOR CHAIR CARRIER SHOULD BE OBTAINED DIRECT FROM MANUFAC-
TURER OF CARRIER.

Figure 13.23 Detail of sink with knee-action valve in Fig. 13.22. (*Courtesy of Crane Plumbing.*)

NORWICH
Vitreous China Lavatory
On Chair Carrier

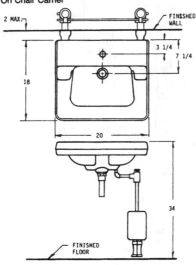

Norwich 1H-164 with Concealed Arms

Lavatory: 1H-164 **Norwich** vitreous china lavatory with rectangular basin, splash lip, front overflow and two soap depressions. Punched for goose-neck spout.

Support:

Trim:

Waste:

Trap:

Supplies:

Color:

Exposed metal trim is polished chromium plated, unless otherwise described.

Figure 13.24   Lavatory with a pedal control. (*Courtesy of Crane Plumbing.*)

NEWCOR
Vitreous China Flushing Rim Service Sink
Siphon Jet Action

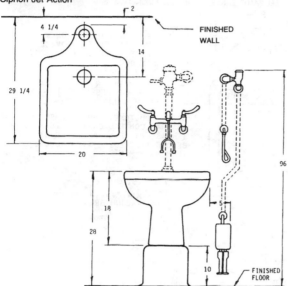

Newcor 7H-538 Flushing Rim
Service Sink Vitreous China

Sink:	7H-538 **Newcor** vitreous china, flushing rim siphon jet service sink; bolt caps.
Trim:	
Pedestal:	
Valve:	
Supply:	
Rim Guard:	
Bedpan Cleanser:	
Note:	For efficient operation of the sink, a minimum flowing water pressure of 20 P.S.I. is required at the valve.

Exposed metal trim is polished chromium plated, unless otherwise described.

Figure 13.25 Flushing rim service sink. (*Courtesy of Crane Plumbing.*)

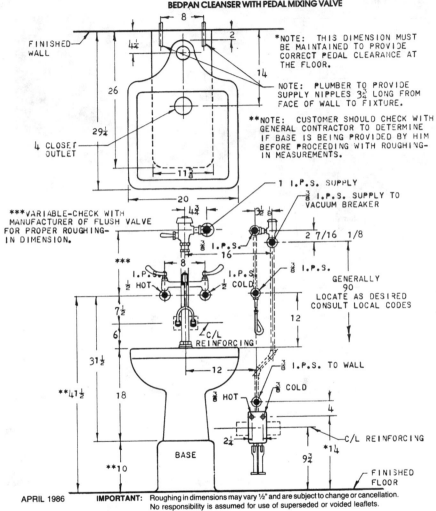

NEWCOR SERVICE SINK
VITREOUS CHINA
DIAL-ESE SUPPLY FITTING
BEDPAN CLEANSER WITH PEDAL MIXING VALVE

*NOTE: THIS DIMENSION MUST BE MAINTAINED TO PROVIDE CORRECT PEDAL CLEARANCE AT THE FLOOR.

NOTE: PLUMBER TO PROVIDE SUPPLY NIPPLES $3\frac{3}{4}$ LONG FROM FACE OF WALL TO FIXTURE.

**NOTE: CUSTOMER SHOULD CHECK WITH GENERAL CONTRACTOR TO DETERMINE IF BASE IS BEING PROVIDED BY HIM BEFORE PROCEEDING WITH ROUGHING-IN MEASUREMENTS.

***VARIABLE-CHECK WITH MANUFACTURER OF FLUSH VALVE FOR PROPER ROUGHING-IN DIMENSION.

APRIL 1986   IMPORTANT: Roughing in dimensions may vary ½" and are subject to change or cancellation. No responsibility is assumed for use of superseded or voided leaflets.

Figure 13.26   Detail of flushing rim service sink in Fig. 13.25. (*Courtesy of Crane Plumbing.*)

WHIRLTON
Siphon Jet Quiet
Action Water Closet

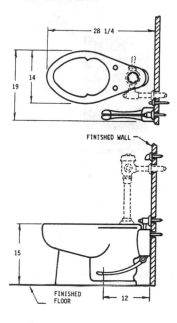

3H-704 Elongated Rim
With Bedpan Lugs

Bowl:	3H-704 **Whirlton** vitreous china water economy, siphon jet, whirlpool quiet action, elongated rim, 1 ½" top spud bowl with integral bedpan lugs; bolt caps
Valve:	
Bedpan Cleanser:	
Seat:	
Color:	
Note:	Bedpan is not included.

Exposed metal trim is polished chromium plated, unless otherwise described.

Figure 13.27   Water closet with bedpan lugs and pedal control. (*Courtesy of Crane Plumbing.*)

SANI-SINK
Vitreous China Flushing Rim Service Sink
Blowout Action

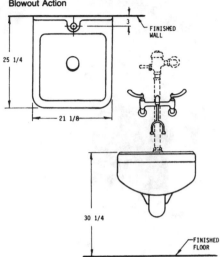

Sani-Sink 7H-544 Flushing Rim
Service Sink Vitreous China

Sink: 7H-544 **Sani-Sink** vitreous China wall-hung flushing rim blowout service sink with 1½" top inlet.

Supports:

Trim:

Valve:

Exposed metal trim is polished chromium plated, unless otherwise described.

Figure 13.28   Flushing rim service sink. (*Courtesy of Crane Plumbing.*)

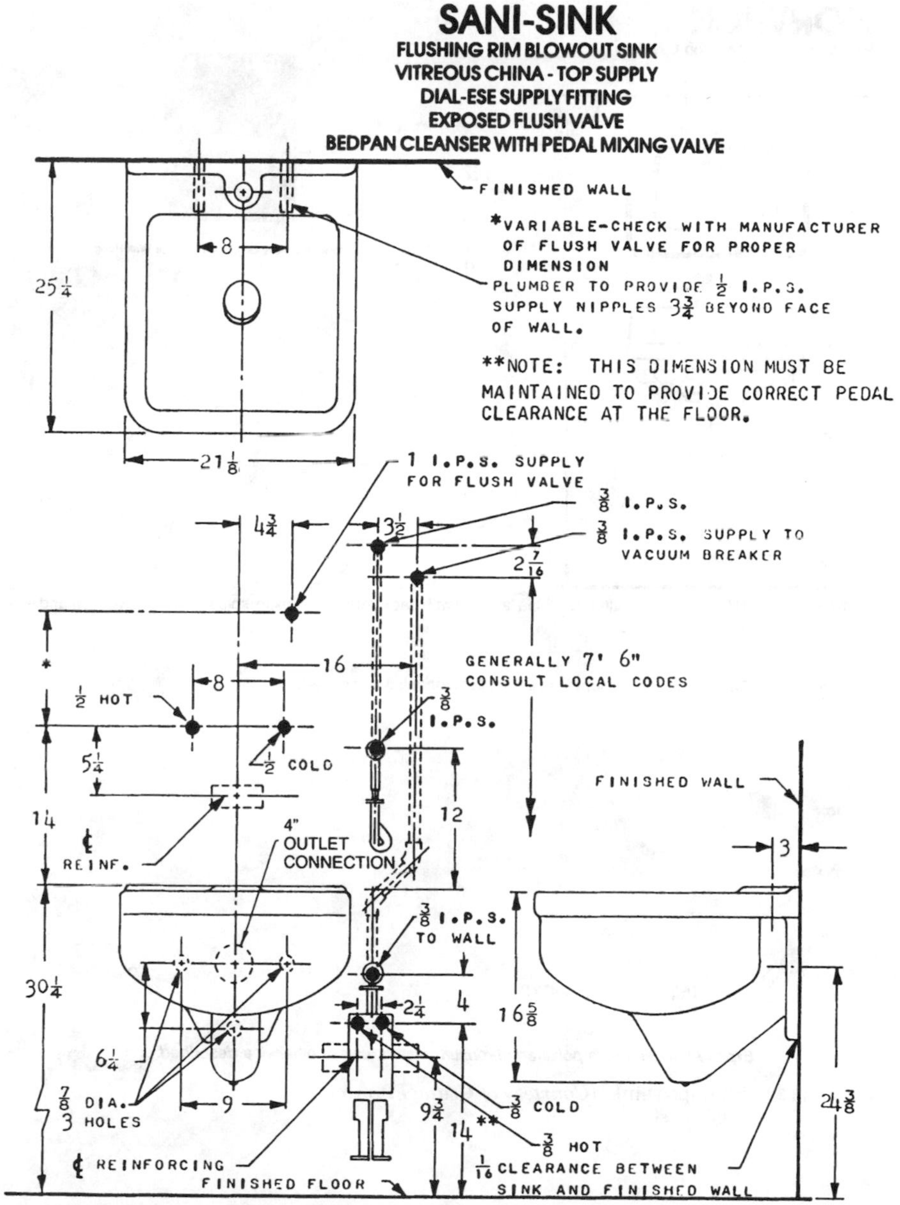

SANI-SINK
FLUSHING RIM BLOWOUT SINK
VITREOUS CHINA - TOP SUPPLY
DIAL-ESE SUPPLY FITTING
EXPOSED FLUSH VALVE
BEDPAN CLEANSER WITH PEDAL MIXING VALVE

Figure 13.29 Detail of flushing rim service sink in Fig. 13.28. (*Courtesy of Crane Plumbing.*)

NORWICH
Vitreous China Shampoo Lavatory

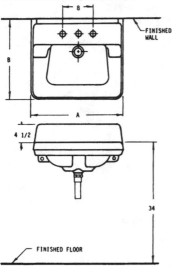

A	B
18	15
20	18
24	21

Norwich 1H-209 On Chair Carrier or Wall Hung

Lavatory: 1H-209 *Norwich* vitreous china lavatory with back, retangular basin, splash lip, front overflow and two soap depressions.

Support: Concealed hanger supplied. Drilled for concealed arm carrier.

Trim:

Trap:

Supplies:

Color:

Size: 18" x 15",   20" x 18",   24" x 21"

Exposed metal trim is polished chromium plated, unless otherwise described.

Figure 13.30   Shampoo sink. (*Courtesy of Crane Plumbing.*)

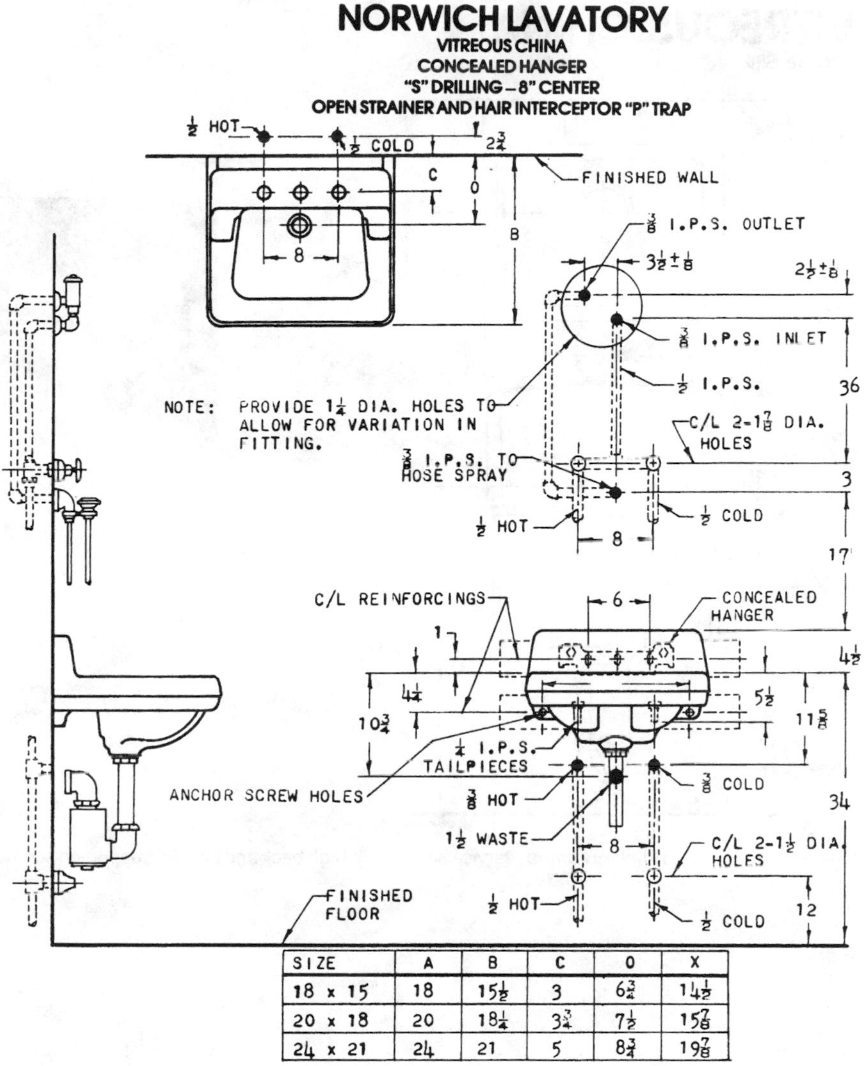
NORWICH LAVATORY
VITREOUS CHINA
CONCEALED HANGER
"S" DRILLING – 8" CENTER
OPEN STRAINER AND HAIR INTERCEPTOR "P" TRAP
1/2 HOT
1/2 COLD
2 3/4
FINISHED WALL
C
O
B
8
3/8 I.P.S. OUTLET
3 1/2 ± 1/8
2 1/2 ± 1/8
3/8 I.P.S. INLET
1/2 I.P.S.
36
NOTE: PROVIDE 1 1/4 DIA. HOLES TO ALLOW FOR VARIATION IN FITTING.
C/L 2-1 7/8 DIA. HOLES
3/8 I.P.S. TO HOSE SPRAY
3
1/2 HOT
1/2 COLD
8
17
C/L REINFORCINGS
6
CONCEALED HANGER
1
4 1/2
4 1/4
5 1/2
10 3/4
11 5/8
1/4 I.P.S. TAILPIECES
ANCHOR SCREW HOLES
3/8 HOT
3/8 COLD
34
1 1/2 WASTE
8
C/L 2-1 1/2 DIA. HOLES
FINISHED FLOOR
1/2 HOT
1/2 COLD
12

SIZE	A	B	C	O	X
18 x 15	18	15 1/2	3	6 3/4	14 1/2
20 x 18	20	18 1/4	3 3/4	7 1/2	15 7/8
24 x 21	24	21	5	8 3/4	19 7/8

Figure 13.31 Lavatory sink. (*Courtesy of Crane Plumbing.*)

VITREOUS CHINA
Service Sink

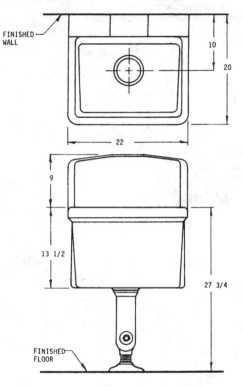

Drilled Back 7-503
Without Rim Guard

Sink: 7-503 ***vitreous china*** , service sink with 8" high back and hole for supporting screw.

Supports:

Trim:

Trap:

Rim Guard:

Exposed metal trim is polished chromium plated, unless otherwise described.

Figure 13.32   Detail of service sink in Fig. 13.31. (*Courtesy of Crane Plumbing.*)

SERVICE SINK
VITREOUS CHINA
DIAL-ESE SUPPLY FITTING
"P" OR "S" TRAP STANDARD
OPTIONAL RIM GUARD

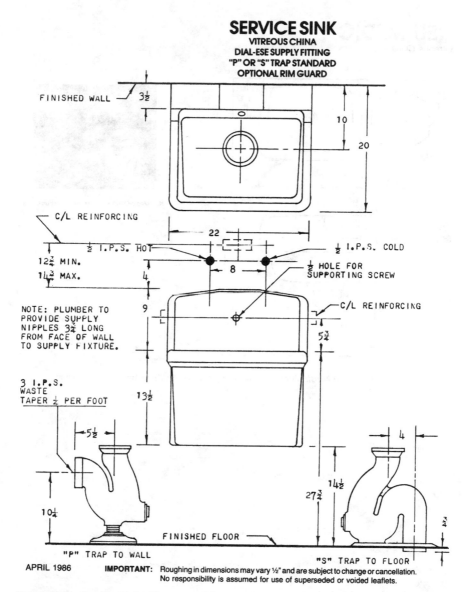

Figure 13.33 Service sink. (*Courtesy of Crane Plumbing.*)

NEU-MEDIC™
Vitreous China Plaster Sink

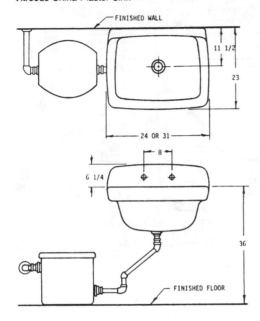

Neu-Medic 5H-258 Plaster Sink Vitreous China

Sink: 5H-258 *Neu-Medic* vitreous china plaster sink with 6¼" high back and return ends.

Supports:

Trim:

Supplies:

Waste:

Trap:

Exposed metal trim is polished chromium plated, unless otherwise described.

Figure 13.34   Detail of plaster sink in Fig. 13.33. (*Courtesy of Crane Plumbing.*)

NEU-MEDIC™ PLASTER SINK
VITREOUS CHINA
DIAL-ESE DOUBLE SUPPLY FITTING
WASTE FITTING WITH OPEN STRAINER
Z-1181 PLASTER INTERCEPTOR

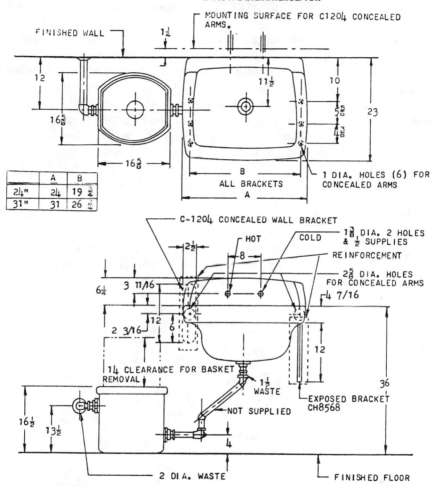

Figure 13.35   Plaster sink. (*Courtesy of Crane Plumbing.*)

Laboratory fittings

Laboratory fittings are very similar to other types of plumbing fittings. There is, however, one very noticeable difference. Laboratory fittings typically have serrated nozzles. This allows the connection of hoses for water, gas, air, and vacuum use. The basic working parts of these fittings are comparable to other types of plumbing fittings.

Other Special Fixtures and Fittings

Other special fixtures and fittings exist in the plumbing world (Figs. 13.30 through 13.35). If you work in commercial kitchens, you are likely to work with prerinse sprayers. While these devices are not common in residential plumbing, they are a way of life in commercial kitchens. However, the troubleshooting skills needed to work with them are no more complex than those used on other types of plumbing.

Whether you are called upon to troubleshoot a garbage can washer, a clinical sink, or a commercial garbage disposer, your basic troubleshooting skills will get you through the job. There will be times when you may need detailed specifications for a given piece of equipment, but the principles and procedures for troubleshooting specialty fixtures are no different from those described throughout this book. Once you have your troubleshooting abilities refined, you can find any type of problem, under any conditions, almost all of the time.

Chapter 14 is going to show you how to troubleshoot well systems. It is a comprehensive chapter with a lot of detailed information on everything from taking meter readings to deciding what symptoms are related to what causes. Let's move onto that chapter now and see what there is to learn about troubleshooting well systems.

14

Troubleshooting Well Systems

Troubleshooting well systems is very difficult for some plumbers. Rural locations depend on wells and water pumps for their potable water, and plumbers moving from a city to a country location can be stymied by the many facets of well systems.

It is not only plumbers from cities who have trouble working with well systems. Those who work with new construction projects have little opportunity to get accustomed to servicing and repairing well systems. Plumbers who are forced into service work by a bad economy and a slowdown in building need a crash course in well systems.

There are three basic types of wells: drilled, driven, and dug wells. Drilled and dug wells are the most common.

As for water pumps, there are two types that are most common: submersible and jet pumps. Jet pumps can work with a one-pipe system or a two-pipe system, depending upon the depth of the well it is serving.

In addition to the well and the water pump, there are many other components involved in a well system. Control boxes, pressure gauges, tank tees, pressure tanks, and air-volume controls are just some of what a plumber has to know about in order to troubleshoot well systems.

A good electrical meter is a necessity for troubleshooting pumps, and a plumber must be able to evaluate meter readings to determine the causes of various types of problems.

Wells

We will begin our education on well systems with the wells themselves. Let's look at each type of well and see how the different types may affect the troubleshooting procedures used.

Driven wells

Driven wells typically are shallow and have small diameters. These wells are used for livestock and occasionally for residential use, but they are not dependable water sources during dry spells. It is not unusual for driven wells to run out of water.

If you are called to a property with a pump that runs and doesn't produce water, the problem may be that the water level has dropped below the pick-up point of the well pipe.

Most driven wells are equipped with a one-pipe, shallow-well jet pump.

Dug wells

Dug wells are much more common for residential uses than driven wells are. In areas with a high water table, such as many areas in the South, dug wells are abundant. These wells have a large diameter and are rarely over 40 feet deep. Many dug wells are less than 30 feet deep.

Modern dug wells are lined with concrete cylinders and capped with heavy concrete tops. This type of well can use a submersible pump, a two-pipe jet pump, or a one-pipe jet pump, depending on the lift of the water. If the lift of water is 25 feet or less, a one-pipe jet pump is normally installed. For deeper wells, either a two-pipe jet pump or a submersible pump is used.

Dug wells are susceptible to pollution and unwanted fill-in conditions. It is not uncommon for mud, sand, pebbles, or other debris to clog the end of a well pipe in a dug well.

Depending upon the depth of the water reservoir, dug wells can run dry during hot, dry months. If you have a pump that is not producing water from a dug well as it should be, check for fill-in conditions and insufficient water depth.

Drilled wells

Drilled wells are the most dependable type of well. Since these wells are drilled deep into the earth, they are usually unaffected by common dry spells.

Drilled wells usually have a modest diameter, a steel casing, and a depth of 100 feet or more. Submersible pumps are the pump of choice for most drilled wells. A two-pipe jet pump can be used with drilled wells with depths of less than 100 feet, but submersible pumps are typically more efficient.

Well-System Components

The number of well-system components a plumber must be familiar with is substantial. Before we dig into the specifics of troubleshooting well systems, let's get acquainted with the many components you may have to deal with.

Shallow-well pumps

Shallow-well pumps use only one pipe to pull water from a well, but the depth of the well should not exceed 25 feet. These pumps work well for shallow wells, but they cannot function when the lift requirements are much more than 25 feet. This type of pump is often called a jet pump. These pumps are installed outside of the well.

Deep-well pumps

Deep-well pumps are very similar in appearance to shallow-well pumps. However, these pumps require a two-pipe system to work properly. One pipe pushes water down into the piping system, and the second pipe pulls the water up to the pressure tank. This type of jet pump is suitable for wells with depths ranging from 25 to 100 feet. These pumps are installed outside of the well.

Submersible pumps

Submersible pumps can be used with any well where the water will cover the pump. Submersible pumps are suspended under the water level in wells.

Rather than pulling water up from the well, submersible pumps push the water up a single pipe to a pressure tank. The design of submersible pumps makes them ideal for deep wells.

The type of pump chosen for a given system is determined largely by the depth of the well. In all cases the pump should not be rated at a higher delivery rate than the recovery rate of the well. If it is, the pump can deplete the water supply in the well. When a well's recovery rate is sufficient, a pump with a 5-gallon-a-minute rating is normally used.

Pressure tanks

Pressure tanks are where water pumped from a well is stored. The use of a pressure tank reduces the wear on a pump and extends its life. Without a pressure tank, a pump would have to cut on and off with each demand for water. The pressure tank allows water to be used to a certain point before the pump must cut on to replenish the water supply.

Most modern pressure tanks are built with an internal diaphragm (Fig. 14.1) to prevent waterlogging. Older tanks don't have this feature and are much more likely to loose their air cushion, resulting in a pump that must cycle more often than it should.

Air-volume controls

Air-volume controls work to maintain the proper volume of air in a pressure tank. Float-type air-volume controls regulate the proper air volume by opening a valve and providing air from the outside atmosphere when the water level rises too high.

Diaphragm-type air-volume controls are more common than float-type controls. These devices allow air to enter the pressure tank each time the pump stops if there is not enough air.

Pressure switches

Pressure switches open and close the electrical circuit for a pump based on preset points. When the pressure in the pressure tank drops to a predetermined point, the pressure switch cuts on and demands water from the pump. As the water pressure reaches the preset limit, the pressure switch cuts the pump off.

Typical settings for pressure switches have them cut on when the pressure drops to 20 pounds per square inch and off when the pressure reaches 40 pounds per square inch. The switch can be set with a cuton of 30 pounds per square inch and a cutoff of 50 pounds per square inch or at any other reasonable settings. There is typically a 20-pound difference between the cutin and the cutout setting.

Pressure gauge

A pressure gauge is used to determine the pressure being produced by the well system (Fig. 14.2).

Relief valve

Relief valves are needed with well systems to prevent excessive pressure buildups. These safety devices should be installed on the discharge side of a pump. As a rule, the relief valve blowoff setting is usually about 20 pounds per square inch higher than the cutout setting on the pressure switch.

1. WELL-X-TROL has a sealed-in air chamber that is pre-pressurized before it leaves our factory. Air and water do not mix eliminating any chance of "waterlogging" through loss of air to system water.

2. When the pump starts, water enters the WELL-X-TROL as system pressure passes the minimum pressure precharge. Only usable water is stored.

3. When the pressure in the chamber reaches maximum system pressure, the pump stops. The WELL-X-TROL is filled to maximum capacity.

4. When water is demanded, pressure in the air chamber forces water into the system. Since WELL-X-TROL does not waterlog and consistently delivers the maximum usable water, minimum pump starts are assured.

Figure 14.1 Diaphragm-type well pressure tank. (*Courtesy of Amtrol, Inc.*)

Figure 14.2   Pressure gauge and pressure switch.

Foot valves

Foot valves act as both a strainer and a check valve. They are installed on the submerged end of suction pipes in wells. The strainer prevents sand and pebbles from entering the pipe, and the check-valve action maintains the prime in the suction line when the pump is not running.

Control boxes

Control boxes (Figs. 14.3 and 14.4) are used with submersible pumps and are the heart of the electrical system. The wiring coming from the pump enters the control box and is routed out to the pressure switch.

Figure 14.3   Control box.

Figure 14.4   Typical deep-well setup of controls.

Other elements of a well system

Other elements of a well system are numerous. There is, of course, the well piping and various fittings, drains, and a multitude of other gadgets and gizmos. We will discuss them as we come to them in our troubleshooting tour.

Troubleshooting Jet Pumps

Troubleshooting jet pumps (Figs. 14.5 through 14.7 and Tables 14.1 and 14.2) is the first hard look we are going to take at pump problems. We will progress in a logical, step-by-step manner that will be easy to understand. Before we dive right into pump problems, let's examine how to use an amprobe and an ohmmeter. These meter readings are essential to the effective troubleshooting of pumps.

Amprobe

An amprobe is a multirange meter that combines the use of an ammeter and a voltmeter. The scales of the voltmeter range from 150 to 600 volts. On the ammeter you will find scales from 5 to 40 amps and from 15 to 100 amps.

The ammeter should be held with its tongs around the wire being tested, with the rotary scale set at the 100-amp range. With this done, rotate the scale back to the smaller numbers until you arrive at an exact reading of the amperage.

When an amprobe is used as a voltmeter, the two leads are connected to the bottom of the meter, with the rotary scale set at 600 volts. When the voltage reading is less than 150 volts, the rotary scale should be set down to 150 volts to obtain a more accurate reading.

Ohmmeter

An ohmmeter measures electrical resistance, and each unit of measure is 1 ohm. If you look at the settings on an ohmmeter, you will notice that there are six of them. The readings are as follows:

- RX1
- RX100
- RX10K
- RX10
- RX1000
- RX100K

The round knob found in the center of an ohmmeter is used to

1. What well conditions might possibly limit the capacity of the pump?	The rate of flow from the source of supply, the diameter of a cased deep well, and the pumping level of the water in a cased deep well.
2. How does the diameter of a cased deep well and pumping level of the water affect the capacity?	They limit the size pumping equipment which can be used.
3. If there are no limiting factors, how is capacity determined?	By the maximum number of outlets or faucets likely to be in use at the same time.
4. What is suction?	A partial vacuum, created in the suction chamber of the pump, obtained by removing pressure due to atmosphere, thereby allowing greater pressure outside to force something (air, gas, water) into the container.
5. What is atmospheric pressure?	The atmosphere surrounding the earth presses against the earth and all objects on it, producing what we call atmospheric pressure.
6. How much is the pressure due to atmosphere?	This pressure varies with elevation or altitude. It is greatest at sea level (14.7 pounds per square inch) and gradually decreases as elevation above sea level is increased. The rate is approximately 1 foot per 100 feet of elevation.
7. What is maximum theoretical suction lift?	Since suction lift is actually that height to which atmospheric pressure will force water into a vacuum, theoretically we can use the maximum amount of this pressure 14.7 pounds per square inch at sea level which will raise water 33.9 feet. From this, we obtain the conversion factor of 1 pound per square inch of pressure equals 2.31-feet head.
8. How does friction loss affect suction conditions?	The resistance of the suction pipe walls to the flow of water uses up part of the work which can be done by atmospheric pressure. Therefore, the amount of loss due to friction in the suction pipe must be added to the vertical elevation which must be overcome, and the total of the two must not exceed 25 feet at sea level. This 25 feet must be reduced 1 foot for every 1000-feet elevation above sea level, which corrects for a lessened atmospheric pressure with increased elevation.
9. When and why do we use a deep-well jet pump?	The resistance of the suction pipe walls to below the pump because this is the maximum practical suction lift which can be obtained with a shallow-well pump at sea level.

Figure 14.5 Questions and answers about pumps. (*Courtesy of A. Y. McDonald Mfg. Co.*)

10. What do we mean by water system?	A pump with all necessary accessories, fittings, etc., necessary for its completely automatic operation.
11. What is the purpose of a foot valve?	It is used on the end of a suction pipe to prevent the water in the system from running back into the source of supply when the pump isn't operating.
12. Name the two basic parts of a jet assembly.	Nozzle and diffuser.
13. What is the function of the nozzle?	The nozzle converts the pressure of the driving water into velocity. The velocity thus created causes a vacuum in the jet assembly or suction chamber.
14. What is the purpose of the diffuser?	The diffuser converts the velocity from the nozzle back into pressure.
15. What do we mean by "driving water"?	That water which is supplied under pressure to drive the jet.
16. What is the source of the driving water?	The driving water is continuously recirculated in a closed system.
17. What is the purpose of the centrifugal pump?	The centrifugal pump provides the energy to circulate the driving water. It also boosts the pressure of the discharged capacity.
18. Where is the jet assembly usually located in a shallow-well jet system?	Bolted to the casing of the centrifugal pump.
19. What is the principal factor which determines if a shallow-well jet system can be used?	Total suction lift.
20. When is a deep-well jet system used?	When the total suction sift exceeds that which can be overcome by atmospheric pressure.
21. Can a foot valve be omitted from a deep-well jet system? Why or why not?	No, because there are no valves in the jet assembly, and the foot valve is necessary to hold water in the system when it is primed. Also, when the centrifugal pump isn't running, the foot valve prevents the water from running back into the well.

Figure 14.6   Questions and answers about pumps. (*Courtesy of A. Y. McDonald Mfg. Co.*)

adjust the meter to zero when the two leads are clipped together. This must be done whenever the range selection for resistance is altered.

It is important to note that an ohmmeter should only be used after the electrical power to the wiring being tested has been turned off.

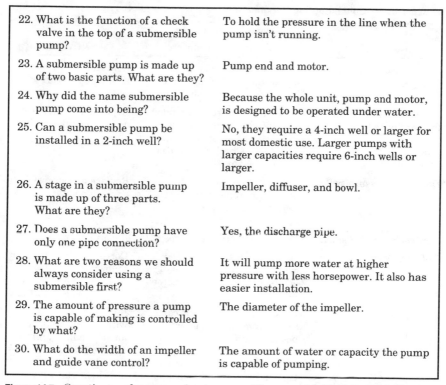

22. What is the function of a check valve in the top of a submersible pump?	To hold the pressure in the line when the pump isn't running.
23. A submersible pump is made up of two basic parts. What are they?	Pump end and motor.
24. Why did the name submersible pump come into being?	Because the whole unit, pump and motor, is designed to be operated under water.
25. Can a submersible pump be installed in a 2-inch well?	No, they require a 4-inch well or larger for most domestic use. Larger pumps with larger capacities require 6-inch wells or larger.
26. A stage in a submersible pump is made up of three parts. What are they?	Impeller, diffuser, and bowl.
27. Does a submersible pump have only one pipe connection?	Yes, the discharge pipe.
28. What are two reasons we should always consider using a submersible first?	It will pump more water at higher pressure with less horsepower. It also has easier installation.
29. The amount of pressure a pump is capable of making is controlled by what?	The diameter of the impeller.
30. What do the width of an impeller and guide vane control?	The amount of water or capacity the pump is capable of pumping.

Figure 14.7 Questions and answers about pumps. (*Courtesy of A. Y. McDonald Mfg. Co.*)

A pump that will not run

A pump that will not run can be suffering from one of many failures. The first thing to check is the fuse or circuit breaker to the circuit. If the fuse is blown, replace it. When the circuit breaker has tripped, reset it.

When the fuse or circuit breaker is not at fault, check for broken or loose wiring connections. Bad connections account for many pump failures.

It is possible the pump won't run because of a motor overload protection device. If the protection contacts are open, the pump will not function. This is usually a temporary condition that corrects itself.

If the pump is attempting to operate at the wrong voltage, it may not run. To test for this condition the power must be on, so be careful. Test the voltage with a volt-ammeter. With the leads attached to the meter and the meter set in the proper voltage range, touch the black lead to the white wire and the red lead to the black wire in the disconnect box near the pump. Test both the incoming and outgoing wiring.

TABLE 14.1   Troubleshooting Suggestions

Cause of trouble	Motor does not start Checking procedure	Corrective action
No power or incorrect voltage.	Using voltmeter, check the line terminals. Voltage must be ± 10% of rated voltage.	Contact power company if voltage is incorrect.
Fuses blown or circuit fuse breakers tripped.	Check fuses for recommended size and check for loose, dirty, or corroded connections in fuse receptacle. Check for tripped circuit breaker.	Replace with proper or reset circuit breaker.
Defective pressure switch.	Check voltage at contact points. Improper contact of switch points can cause voltage less than line voltage.	Replace pressure switch or clean points.
Control box malfunction.	For detailed procedure, see ***	Repair or replace.
Defective wiring.	Check for loose or corroded connections. Check motor ead terminals with voltmeter lfor power.	Correct faulty wiring or connections.
Bound pump.	Locked rotor conditions can result from misalignment between pump and motor or a sand bound pump. Amp readings 3 to 6 times higher than normal will be indicated.	If pump will not start with several trials, it must be pulled and the cause corrected. New installations should always be run without turning off until water clears.
Defective cable or motor.	For detailed procedure, see pp. 291, 294, and 295.	Repair or replace.
	Motor starts too often	
Pressure switch.	Check setting on pressure switch and examine for defects.	Reset limit or replace switch.
Check valve, stuck open.	Damaged or defective check valve will not hold pressure.	Replace if defective.
Waterlogged tank (air supply).	Check air-charging system for proper operation.	Clean or replace.
Leak in system.	Check system for leaks.	Replace damaged pipes or repair leaks.

SOURCE: A. Y. McDonald Manufacturing Co.

Your next step in the testing process should be at the pressure switch. The black lead should be placed on the black wire, and the red lead should be put on the white wire for this test. There should be a plate on the pump that identifies the proper working voltage. Your

TABLE 14.2   Troubleshooting Suggestions

Causes of trouble	Motor runs continuously Checking procedure	Corrective action
Pressure switch.	Switch contacts may be "welded" in closed position. Pressure switch may be set too high.	Clean contacts, replace switch, or readjust setting.
Low-level well.	Pump may exceed well capacity. Shut off pump, wait for well to recover. Check static and drawdown level from well head.	Throttle pump output or reset pump to lower level. Do not lower if sand may clog pump.
Leak in system.	Check system for leaks.	Replace damaged pipes or repair leaks.
Worn pump.	Symptoms of worn pump are similar to that of drop pipe leak or low water level in well. Reduce pressure switch setting. If pump shuts off, worn parts may be at fault. Sand is usually present in tank.	Pull pump and replace worn impellers, casing, or other close fitting parts.
Loose or broken motor shaft.	No or little water will be delivered if coupling between motor and pump shaft is loose or if a jammed pump has caused the motor shaft to shear off.	Check for damaged shafts if coupling is loose, and replace worn or defective units.
Pump screen blocked.	Restricted flow may indicate a clogged intake screen on pump. Pump may be installed in mud or sand.	Clean screen and reset at less depth. It may be necessary to clean well.
Check valve stuck closed.	No water will be delivered if check valve is in closed position.	Replace if defective.
Control box malfunction.	See pages 296, 297, and 305 for single phase.	Repair or replace.
	Motor runs but overload protector tips	
Incorrect voltage.	Using voltmeter, check the line terminals. Voltage must be within ± 10% of rated voltage.	Contact power company if voltage is incorrect.
Overheated protectors.	Direct sunlight or other heat source can make control box hot, causing protectors to trip. The box must not be hot to touch.	Shade box, provide ventilation, or move box away from heat source.
Defective control box.	For detailed procedures, see pages 295, 300, and 301.	Repair or replace.
Defective motor or cable.	For detailed procedures, see pages 294, 301, and 304.	Repair or replace.
Worn pump or motor.	Check running current. See pages 294, 305, and 306.	Replace pump and/or motor.

SOURCE: A. Y. McDonald Manufacturing Co.

test should reveal voltage that is within 10 percent of the recommended rating.

An additional problem that you may encounter is a pump that is mechanically bound. You can check this by removing the end cap and turning the motor shaft by hand. It should rotate freely.

A bad pressure switch can keep a pump from running. Remove the cover from the pressure switch, and you will see two springs, one tall and one short. These springs are depressed and held in place by individual nuts.

The short spring is preset at the factory and should not need adjustment. This adjustment controls the cutout sequence for the pump. If you turn the nut down, the cutout pressure will be increased. Loosening the nut will lower the cutout pressure.

The long spring can be adjusted to change the cutin and cutout pressure for the pump. If you want to set a higher cutin pressure, turn the nut tighter to depress the spring further. To reduce the cutin pressure you should loosen the nut to allow more height on the spring. If the pressure switch fails to respond to the adjustments, it should be replaced.

It is also possible that the tubing or fittings on the pressure switch are plugged. Take the tubing and fittings apart and inspect them. Remove any obstructions and reinstall them.

The last possibility for the pump failure is a bad motor. Use an ohmmeter to check the motor after the power to the pump has been turned off.

Start checking the motor by disconnecting the motor leads. We will call these leads L1 and L2. The instructions you are about to receive are for Goulds pumps with motors rated at 230 volts. When you are conducting the test on different types of pumps, you should refer to the manufacturer's recommendations.

Set the ohmmeter to RX100 and adjust the meter to zero. Put one of the meter's leads on a ground screw. The other lead should be systematically touched to all terminals on the terminal board, switch, capacitor, and protector. If the needle on your ohmmeter doesn't move as these tests are made, the ground check of the motor is okay.

The next check to be conducted is for winding continuity. Set the ohmmeter to RX1 and adjust it to zero. You will need a thick piece of paper for this test; it should be placed between the motor switch points and the discharge capacitor. You should read the resistance between L1 and A to see that it is the same as the resistance between A and yellow. The reading between yellow to red should be the same as L1 to the same red terminal.

The next test is for the contact points of the switch. Set the ohmmeter to RX1 and adjust it to zero. Remove the leads from the switch and attach the meter leads to each side of the switch; you should see a reading of zero. If you flip the governor weight to the run position, the reading on your meter should be infinity.

Now let's check the overload protector. Set your meter to RX1 and adjust it to zero. With the overload leads disconnected, check the resistance between terminals 1 and 2 and then between 2 and 3. If a reading of more than 1 occurs, replace the overload protector.

The capacitor can also be tested with an ohmmeter. Set the meter to RX1000 and adjust it to zero. With the leads disconnected from the capacitor, attach the meter leads to each terminal. When you do this, you should see the meter's needle go to the right and drift slowly to the left. To confirm your reading, switch positions with the meter leads and see if you get the same results. A reading that moves toward zero or a needle that doesn't move at all indicates a bad capacitor.

Pump runs but gives no water

When a pump runs but gives no water, you have seven possible problems to check out. Let's take a look at each troubleshooting phase in its logical order.

The first consideration should be that of the pump's prime. If the pump or the pump's pipes are not completely primed, water will not be delivered.

For a shallow-well pump you should remove the priming plug and fill the pump completely with water. You may want to disconnect the well pipe at the pump and make sure it is holding water. You could spend considerable time pouring water into a priming hole only to find out the pipe was not holding the water.

For deep-well jet pumps, you must check the pressure-control valves. The setting must match the horsepower and jet assembly used, so refer to the manufacturer's recommendations.

Turning the adjustment screw to the left will reduce pressure, and turning it to the right will increase pressure. When the pressure-control valve is set too high, the air-volume control cannot work. If the pressure setting is too low, the pump may shut itself off.

If the foot valve or the end of the suction pipe has become obstructed or is suspended above the water level, the pump cannot produce water. Sometimes shaking the suction pipe will clear the foot valve and get the pump back into normal operation. If you are working with a two-pipe system, you will have to pull the pipes and do a visual inspection. However, if the pump you are working on is a one-pipe pump, you can use a vacuum gauge to determine if the suction pipe is blocked.

If you install a vacuum gauge in the shallow-well adapter on the pump, you can take a suction reading. When the pump is running, the gauge will not register any vacuum if the end of the pipe is not below the water level or if there is a leak in the suction pipe.

An extremely high vacuum reading, such as 22 inches or more, indicates that the end of the pipe or the foot valve is blocked or buried in mud. It can also indicate that the suction lift exceeds the capabilities of the pump.

A common problem when the pump runs without delivering water is a leak on the suction side of the pump. You can pressurize the system and inspect it for these leaks.

The air-volume control can be at fault for a pump that runs dry. If you disconnect the tubing and plug the hole in the pump, you can tell if the air-volume control has a punctured diaphragm. If plugging the pump corrects the problem, you must replace the air-volume control.

Sometimes the jet assembly will become plugged up. When this happens with a shallow-well pump, you can insert a wire through the ½-inch plug in the shallow-well adapter to clear the obstruction. With a deep-well jet pump, you must pull the piping out of the well and clean the jet assembly.

An incorrect nozzle or diffuser combination can result in a pump that runs but that produces no water. Check the ratings in the manufacturer's literature to be sure the existing equipment matches them.

The foot valve or an in-line check valve could be stuck in the closed position. This type of situation requires a physical inspection and the probable replacement of the faulty part.

Pump cycles too often

When a pump cycles on and off too often, it can wear itself out prematurely. This type of problem can be caused by several things. For example, leaks in the piping or pressure tank would cause frequent cycling of the pump.

The pressure switch may be responsible for a pump that cuts on and off to often. If the cutin setting on the pressure gauge is set too high, the pump will work harder than it should.

If the pressure tank becomes waterlogged (filled with too much water and not enough air), the pump will cycle frequently. If the tank is waterlogged, it will have to be recharged with air. This would also lead you to suspect that the air-volume control is defective.

An insufficient vacuum could cause the pump to run too often. If the vacuum does not hold at 3 inches for 15 seconds, it might be the problem.

The last thing to consider is the suction lift. It's possible that the pump is getting too much water and creating a flooded suction. This can be remedied by installing and partially closing a valve in the suction pipe.

Won't develop pressure

Sometimes a pump will produce water but will not build the desired pressure in the holding tank. Leaks in the piping or pressure tank can cause this condition.

If the jet or the screen on the foot valve is partially obstructed, the same problem may result.

A defective air-volume control may prevent the pump from building suitable pressure. You can test for this by removing the air-volume control and plugging the hole where it was removed. If this solves the problem, you know the air-volume control is bad.

A worn impeller hub or guide vane bore could result in a pump that will not build enough pressure. The proper clearance should be 0.012 on a side, or 0.025 diametrically.

With a shallow-well system, the problem could be being caused by the suction lift being too high. You can test for this with a vacuum gauge. The vacuum should not exceed 22 inches at sea level. For deep-well jet pumps you must check the rating tables to establish their maximum jet depth. Also with deep-well jet pumps, you should check the pressure-control valve to see that it is set properly.

Switch fails to cut out

If the pressure switch fails to cut out when the pump has developed sufficient pressure, you should check the settings on the pressure switch. Adjust the nut on the short spring and see if the switch responds; if it doesn't, replace the switch.

Another cause for this type of problem could be debris in the tubing or fittings between the switch and pump. Disconnect the tubing and fittings and inspect them for obstructions.

We have now covered the troubleshooting steps for jet pumps, but before we move on to submersible pumps, look over the illustrations I've given you that show engineering data, multistage jet pumps, and typical installation procedures (Figs. 14.8 through 14.12).

Troubleshooting Submersible Pumps

There are some major differences between troubleshooting submersible pumps and jet pumps. One of the most obvious differences is that jet pumps are installed outside of wells, and submersible pumps are installed below the water level of wells (Fig. 14.13).

WX-100 Series WELL-X-TROL with tank mounted at jet pump.

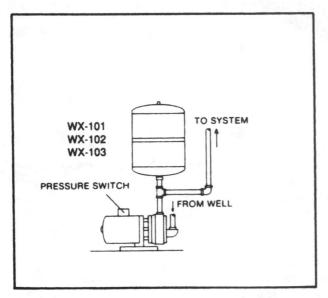

Figure 14.8 Pressure tank mounted at jet pump. (*Courtesy of Amtrol, Inc.*)

WX-100 Series WELL-X-TROL installed on-line with jet pump.

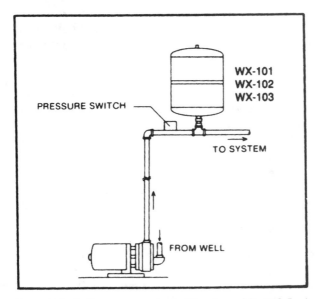

Figure 14.9 In-line pressure tank. (*Courtesy of Amtrol, Inc.*)

**WX-103 or WX-200 with #162 pump stand.
WX-200 with #163 pump stand mounting jet
pump.**

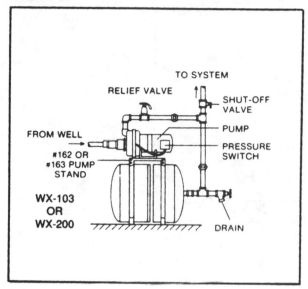

Figure 14.10 Jet pump mounted on pressure tank. (*Courtesy of Amtrol, Inc.*)

There are times when a submersible pump must be pulled out of a well, and this can be quite a chore. Even with today's lightweight well pipe, the strength and endurance needed to pull a submersible pump up from a deep well is considerable. Plumbers who work with submersible pumps regularly often have a pump puller to make removing the pumps easier.

When a submersible pump is pulled, you must allow for the length of the well pipe when planning the direction to pull from and where the pipe and pump will lay once removed from the well. It is not unusual to have between 100 and 200 feet of well pipe to deal with, and some wells are even deeper.

It is important when pulling a pump, or lowering one back into a well, that the electrical wiring does not rub against the well casing. If the insulation on the wiring is cut, the pump will not work properly. Let's look now at some specific troubleshooting situations.

Pumps that won't start

Pumps that won't start may be victims of a blown fuse or tripped circuit breaker. If these conditions are okay, turn your attention to the voltage.

WX-201 through WX-350 WELL-X-TROL installed on-line with jet pump.

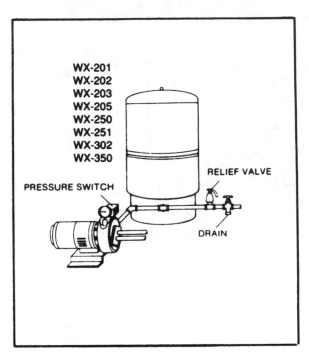

WX-201
WX-202
WX-203
WX-205
WX-250
WX-251
WX-302
WX-350

RELIEF VALVE

PRESSURE SWITCH

DRAIN

Figure 14.11 Jet pump mounted on floor, next to pressure tank. (*Courtesy of Amtrol, Inc.*)

In the following scenarios we will be dealing with Goulds pumps and Q-D-type control boxes.

To check the voltage, remove the cover of the control box to break all motor connections. Be advised: Wires L1 and L2 are still connected to electrical power. These are the wires running to the control box from the power source.

Press the red lead from your voltmeter to the white wire and the black lead to the black wire. Keep in mind that any major electrical appliance that might be running at the same time the pump would be, like a clothes dryer, should be turned on while you are conducting your voltage test.

Once you have a voltage reading, compare it to the manufacturer's recommended ratings. For example, with a Goulds pump that is rated for 115 volts, the measured volts should range from 105 to 125. A pump with a rating of 208 volts should measure a range from 188 to

WX-201 through WX-203 using #161 pump stand. WX-205 through WX-350 using #165 pump stand. Shallow well jet pump mounted on tank.

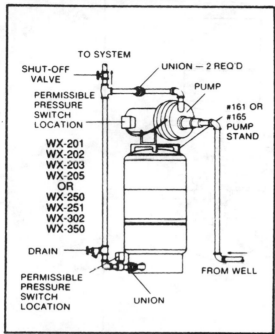

Figure 14.12   Jet pump mounted on pressure tank. *(Courtesy of Amtrol, Inc.)*

228 volts. A pump rated at 230 volts should measure between 210 and 250 volts.

If the voltage is okay, check the points on the pressure switch. If the switch is defective, replace it.

The third likely cause of this condition is a loose electrical connection in the control box, the cable, or the motor. Troubleshooting for this condition requires extensive work with your meters.

To begin the electrical troubleshooting, we will look for electrical shorts by measuring the insulation resistance. Use an ohmmeter for this test; the power to the wires you are testing should be turned off.

Set the ohmmeter scale to RX100K and adjust it to zero. You will be testing the wires coming out of the well, from the pump, at the well head. Put one of the ohmmeter's lead to any one of the pump wires and place the other ohmmeter lead on the well casing or a

WX-201 through WX-350 WELL-X-TROL installed on-line using submersible pump.

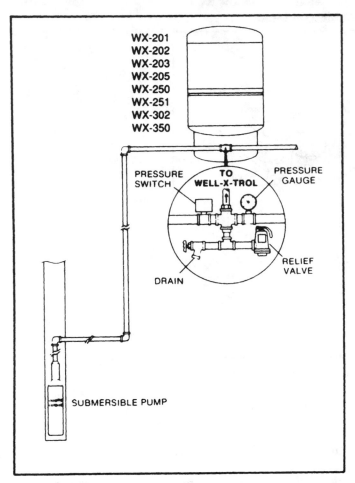

Figure 14.13   In-line pressure tank with submersible pump. (*Courtesy of Amtrol, Inc.*)

metal pipe. As you test the wires for resistance, you will need to know what the various readings mean, so let's examine this issue.

You will be dealing with normal ohm and megohm values. Insulation resistance does not vary with ratings. Regardless of the motor, horsepower, voltage, or phase rating, the insulation resistance remains the same.

A new motor that has not been installed should have an ohm value of 20,000,000 or more and a megohm value of 20. A motor that has been used but is capable of being reinstalled should produce an ohm reading of 10,000,000 or more and a megohm reading of 10.

For a motor that is installed in the well, which will be the case in most troubleshooting, the readings will be different. A new motor installed with its drop cable should give an ohm reading of 2,000,000 or more and a megohm value of 2. An installed motor in a well that is not new but is in good condition will present an ohm reading of between 500,000 and 2,000,000. Its megohm value will be between 0.5 and 2.

A motor that gives a reading in ohms of between 20,000 and 500,000 and a megohm reading of between 0.02 and 0.5 may have damaged leads or may have been hit by lightning; however, don't pull the pump yet.

You should pull the pump when the ohm reading ranges from 10,000 to 20,000 and the megohm value drops to between 0.01 and 0.02. These readings indicate a motor that is damaged or cables that are damaged. While a motor in this condition may run, it probably won't run for long.

When a motor has failed completely or the insulation on the cables has been destroyed, the ohm reading will be less than 10,000 and the megohm value will be between 0 and 0.01.

With this phase of the electrical troubleshooting done, we are ready to check the winding resistance. You will have to refer to charts as reference for correct resistance values, and you will have to make adjustments if you are reading the resistance through the drop cables. I'll explain more about this in a moment.

If the ohm value is normal during your test, the motor windings are not grounded and the cable insulation is intact. When the ohm readings are below normal, you will have discovered that either the insulation on the cables is damaged or the motor windings are grounded.

To measure winding resistance with the pump still installed in the well, you will have to allow for the size and length of the drop cable. Assuming you are working with copper wire, you can use the following figures to obtain the resistance of cable for each 100 feet in length and ohms per pair of leads:

Cable size	Resistance
14	0.5150
12	0.3238
10	0.2036
8	0.1281
6	0.08056
4	0.0506
2	0.0318

If aluminum wire is being tested, the readings will be higher. Divide the ohm readings above by 0.61 to determine the actual resistance of aluminum wiring.

If you pull the pump and check the resistance for the motor only (not with the drop cables being tested), you will use different ratings. You should refer to a chart supplied by the manufacturer of the motor for the proper ratings.

When all the ohm readings are normal, the motor windings are fine. If any of the ohm values are below normal, the motor is shorted. An ohm value that is higher than normal indicates that the winding or cable is open or that there is a poor cable joint or connection. Should you encounter some ohm values that are higher than normal while others are lower than normal, you have found a situation where the motor leads are mixed up and need to be attached in their proper order.

If you want to check an electrical cable or a cable splice, you will need to disconnect the cable and have a container of water to submerge it in; a bathtub will work. Start by submerging the entire cable, except for the two ends, in water. Set your ohmmeter to RX100K and adjust it to zero. Put one of the meter leads on a cable wire and the other to a ground. Test each wire in the cable with this same procedure.

If at any time the meter's needle goes to zero, remove the splice connection from the water and watch the needle. If the needle falls back to give no reading, the leak is in the splice.

Once the splice is ruled out, you have to test sections of the cable in a similar manner. In other words, once you have activity on the meter, you should slowly remove sections of the cable until the meter settles back into a no-reading position. When this happens, you have found the section that is defective. At this point, the leak can be covered with waterproof electrical tape and reinstalled, or you can replace the cable.

A pump that will not run

A pump that will not run can require extensive troubleshooting. Start with the obvious and make sure the fuse is not blown and the circuit breaker is not tripped. Also check to see that the fuse is of the proper size.

Incorrect voltage can cause a pump to fail. You can check the voltage as described in the electrical troubleshooting section above.

Loose connections, damaged cable insulation, and bad splices, as discussed above, can prevent a pump from running.

The control box can have a lot of influence on whether or not a pump will run. If the wrong control box has been installed or if the box is located in an area where temperatures rise to over 122 degrees F, the pump may not run.

When a pump will not run, you should check the control box out carefully. We will be working with a quick-disconnect-type box. Start by checking the capacitor with an ohmmeter. First, discharge the capacitor before testing. You can do this by putting the metal end of a screwdriver between the capacitor's clips. Set the meter to RX1000 and connect the leads to the black and orange wires out of the capacitor case. You should see the needle start toward zero and then swing back to infinity. Should you have to recheck the capacitor, reverse the ohmmeter leads.

The next check involves the relay coil. If the box has a potential relay (three terminals), set your meter on RX1000 and connect the leads to red and yellow wires. The reading should be between 700 and 1800 ohms for 115-volt boxes. A 230-volt box should read between 4500 and 7000 ohms.

If the box has a current relay coil (four terminals), set the meter on RX1 and connect the leads to black wires at terminals 1 and 3. The reading should be less than 1 ohm.

In order to check the contact points, set your meter on RX1 and connect to the orange and red wires in a three-terminal box. The reading should be zero. For a four-terminal box, set the meter at RX1000 and connect to the orange and red wires. The reading should be near infinity.

Now you are ready to check the overload protector with your ohmmeter. Set the meter at RX1, and connect the leads to the black wire and the blue wire. The reading should be at a maximum of 0.5.

If you are checking the overload protector for a control box designed for 1½ horsepower or more, set your meter at RX1 and connect the leads to terminals 1 and 3 on each overload protector. The maximum reading should not exceed 0.5 ohms.

A defective pressure switch or an obstruction in the tubing and fittings for the pressure switch could cause the pump not to run.

As a final option, the pump may have to be pulled and checked to see if it is bound. There should be a high amperage reading if this is the case.

Pump runs but doesn't produce water

When a submersible pump runs but doesn't produce water, there are several things that could be wrong. The first thing to determine is if the pump is submerged in water. If you find that the pump is submerged, you must begin your regular troubleshooting.

Loose connections or wires connected incorrectly in the control box could be at fault. The problem could be related to the voltage. A leak in the piping system could easily cause the pump to run without producing adequate water. A check valve could be stuck in the closed position. If the pump was just installed, the check valve may be installed backward.

Other options include a worn pump or motor, a clogged suction screen or impeller, and a broken pump shaft or coupling. You will have to pull the pump if any of these options are suspected.

Enough tank pressure

If you don't have enough tank pressure, check the setting on the pressure switch. If that's okay, check the voltage. Next, check for leaks in the piping system, and as a last resort, check the pump for excessive wear.

Frequent cycling

Frequent cycling is often caused by a waterlogged tank, as was described in the section on jet pumps. Of course, an improper setting on the pressure switch can cause a pump to cut on too often, and leaks in the piping can be responsible for the trouble.

You may find the problem is being caused by a check valve that has stuck in an open position. Occasionally the pressure tank will be sized improperly and cause problems. The tank should allow a minimum of 1 minute of running time for each cycle.

Get Familiar with Your Meters

Before you attempt to troubleshoot pumps, get familiar with your meters. Meter readings play a vital role in the successful troubleshooting of pumps, so you must know how to take, read, and interpret your meter.

Once you master your meter, you are well on your way to being a good troubleshooter for pump problems. Ideally, you should equip yourself with charts and tables from various pump manufacturers for reference in the field.

With water pumps behind us, let's move on to the next chapter and learn about septic systems.

15

Troubleshooting Septic Systems

Troubleshooting septic systems is not a task that many urban plumbers are faced with. Rural plumbers, on the other hand, are frequently called for problems that are related to septic systems. The problems range from backed up drains to foul odors. If you work in an area where septic systems are used, you must have a working knowledge of how these waste-disposal systems work.

Most plumbers who work in areas with septic systems know a little about their design and layout, but it is surprising how little most plumbers know about the actions of a septic system.

Plumbers don't usually install or maintain septic systems. They are generally only responsible for the connection between the septic tank and the building drain. This would seem to remove septic systems from the concerns of plumbers, but real life doesn't make the break so easily.

When someone flushes a toilet and it backs up into their bathtub, that person is probably going to call a plumber. The problem may well lie in the septic tank, but a panicked property owner is going to call a plumbing company.

A plumber who is not familiar with septic systems could spend all day working to clear a stoppage in the building drain or sewer with no success. Plumbers experienced with septic systems will look to the septic tank early in the troubleshooting process.

A homeowner who is smelling a foul odor will often call in a plumbing professional. The smell of sewage would lead many plumbers to look for dry traps or improper venting. However, if the house is served by a septic system that is overloaded with liquids, the smell may be coming from the drainfield. I had this happen not too long ago with one of my plumbers.

We got a call from a homeowner who thought she might have a broken drainage pipe under her home. The property owner could smell what seemed to be sewage in the early morning of most days.

The plumber dispatched to the call was young and not very experienced in septic systems. In my work area, some houses are on municipal sewers and others are on private waste disposal systems.

When the plumber arrived at the home, he could smell the acrid odor. Being young and inexperienced, he wasn't sure of what to look for. He walked all through the house and noticed that the smell was most noticeable in the hallway. There happened to be an open French door at the end of the hall, but this didn't tip the plumber off on the possibility of a flooded drainfield.

After about half an hour of searching through the house, my plumber told the homeowner that he believed the problem was a leak in a vent. He had checked all the fixture traps, and they were sealed properly.

When the homeowner asked what the next step in solving the problem would be, the plumber indicated that he would have to start opening walls to find the location of the leak. The homeowner panicked and called my office. When the receptionist put her call through to me, the lady was hysterical.

I let her talk herself out, and then I explained to her that the plumber was wrong in his troubleshooting decision. Deciding I had a serious customer situation on my hands, I said that I would come out to the job personally and see what was wrong.

I couldn't believe that the plumber had told the homeowner we would have to go through her house cutting walls open at random to locate the problem. Even if the problem had been with the venting system, cutting walls open at random would not have been proper procedure. A smoke or peppermint test would have revealed the leak in the venting system without destroying the house.

When I arrived at the house, I went immediately to the distraught homeowner. I talked with her and she calmed down. Then I went to my plumber and told him to go back to the office and wait for me; I was going to want to talk to him in private about his way of handling the job.

By the time I arrived at the job, the odor was fading. The homeowner explained that it was at its worst early in the morning and that by the afternoon it was tolerable. This in itself told me that we were not dealing with an interior plumbing problem. However, to be safe, I checked all of the traps, and they were sealed.

I looked in the hallway and saw the open French door. Outside the door was a large, lush lawn. There were no vent pipes in the hall walls, but the heavy green grass indicated a strong possibility of a drainage field from a septic tank.

I asked the homeowner if her house was connected to a septic system. She said it was. When asked where the system was located, the lady didn't know. I went out into the backyard, and the ground was so saturated; it squished with every step. As the moisture in the ground was disturbed, the unpleasant smell of a septic system rose.

The property owner's leech field was flooded. The early morning dampness made the odor linger until the days warmed and the odors could rise. The open French door and the stairs to the upper level of the house acted as a chimney to pull the odor into the home.

What my plumber thought was a plumbing problem was actually a drainfield problem. The woman's fear of having her house cut open was eliminated quickly with my knowledge of septic systems. The homeowner still had a serious problem, but I had avoided a problem for my company by not wrecking the house only to come up empty.

This example should prove to you that plumbers, even those who rarely work with septic systems, do need a basic understanding of how the systems work and why they sometimes don't.

Septic-System Components

What are the components of a septic system? The two main parts are a septic tank and a drainfield, but there is more involved in the makeup of a working septic system. Let's look, one at a time, at the components that may be found in a common septic system.

Septic tanks

Septic tanks are the first element needed for a successful private waste-disposal system. These tanks must be waterproof, and they receive the discharge from a building's sewer.

Most modern septic tanks are made of concrete, but there are still many tanks in operation that are made of metal. The metal tanks rust out in time, and this causes problems for property owners and plumbers.

A septic tank is designed to separate liquid waste from solid waste. An average house has a septic tank with a capacity of 1000 gallons. Many of the old metal tanks only hold 500 gallons. In most cases, a 750-gallon tank is the smallest that should be used with a residential property. In theory, a septic tank should be large enough to hold the capacity of discharge from a building based on a 24-hour usage. This formula is meant to allow the tank time to break down the solids.

Septic tanks are generally regulated by the local health department or an environmental agency. Commercial septic tanks typically range in size from 750 to 3600 gallons.

The sewer pipe running from a building to the inlet of a septic tank should be a solid pipe or an approved material, such as PVC, ABS, or cast iron.

Distribution boxes

Distribution boxes are found on the outlet side of septic tanks. These boxes are usually made of concrete, and they collect liquids coming from the septic tank and disperse them into individual drainfield lines.

The pipe running from a septic tank to a distribution box should be solid and of an approved material, such as PVC, ABS, or cast iron.

The pipe from the distribution box to the individual disposal lines is normally slotted, plastic pipe.

Drainfields

Drainfields can be made in many configurations. The design of a drainfield is usually done by a private engineer or a local official. An average drainfield contains slotted pipe, but some drainfields, in areas where percolation is poor, are be made with chamber systems.

When slotted pipe is used, it is typically installed with a maximum grade of $\frac{1}{2}$ inch for each 10 liner feet of pipe. The pipe is laid on a bed of $\frac{3}{4}$-inch washed stone in most instances. The stone is normally 6 inches deep below the pipe and 6 inches deep above the pipe, for a total depth of 12 inches.

Siphon chambers

Siphon chambers are not normally used, but when they are, they are installed between the septic tank and the distribution box. The pipe on both sides of a siphon chamber should be solid and of an approved material.

Siphon chambers are used to regulate the flow of liquids into the drainfield. By creating a sudden flow of liquid, from siphonic action, the siphon chamber enables all drain lines to receive adequate liquids at about the same time. These chambers improve the efficiency of a waste-disposal system, but they are not a required feature.

Grease intercepters

Grease intercepters, or grease traps as they are often called, are sometimes installed on a drainage line en route to a septic tank. When this is the case, the grease intercepter is normally installed on the drain of a kitchen sink as it heads for the main sewer, outside of the building. Many grease intercepters, however, are installed within the confines of a building.

Grease must be kept out of a septic system or the ability of the septic tank to break down materials will be hindered. If grease makes its way to the slotted pipes of a drainfield, it can clog the pipes and render the system useless.

How Does a Septic System Work?

Few plumbers can tell you how a septic system works. While they often know that a septic system separates solids from liquids and disposes of the liquids, not many plumbers know how this feat is accomplished. Since the effectiveness of a septic tanks relies on what comes into it, plumbers should know what goes on in the tanks, and that is what you are about to learn.

In simple terms, solids sink to the bottom of a septic tank and liquids are passed on to the drainage field. This process, however, is not as simple as it sounds.

Bacteria, known as anaerobic bacteria, works to break down the contents of a septic tank into manageable sizes. Anaerobic bacteria thrives in areas where oxygen is not present, such as in a covered septic tank.

Sludge is the accumulation of solids on the bottom of a septic tank. It is gradually broken down into smaller pieces, but there will always be some sludge buildup on the bottom of the tank. When a septic tank is pumped out, the sludge is usually dislodged and sucked into the tanker truck.

Scum exists in septic tanks, and it floats, partially submerged, near the top of the tank. Made up of floating solids, scum will continue to break down to some extent. Grease, fat, and similar materials contribute to the scum layer.

Measuring the depth of scum and sludge is the proper way to determine when a septic tank should be pumped clean. If the bottom of the scum is within 3 inches of the outlet invert, the tank should be pumped. If the scum gets too close to the outlet, scum may be transported into the drainfield and cause serious problems.

A 1000-gallon septic tank with a 4-foot liquid level should have a clear distance of at least 6 inches between the sludge layer and the outlet invert. If the sludge is closer than 6 inches to the outlet, the tank should be pumped.

The effluent (liquid passing out of the septic tank into the septic field) is not pure or safe. It is rendered harmless in the drainfield, not in the septic tank. Once the effluent has seeped through an approved septic field, it is safe to enter the groundwater supply.

Troubleshooting Septic Systems

Troubleshooting septic systems can be a smelly job, but somebody has to do it. What do you, as a plumber, have to do to troubleshoot a septic system? Mostly, you only have to know how to determine if a building's plumbing problem is being caused by a faulty septic system. You don't necessarily have to know how to correct the septic problem, but you do need to know how to identify it.

In the example I gave earlier about one of my plumbers and the flooded drainfield, you could see how a lack of knowledge pertaining to the operation of a private waste-disposal system can affect a plumber. If I had not intervened, the plumber would probably have chopped open a bunch of walls and been sued for having made a false diagnosis. Whether you are working for an employer or are self-employed, you don't want to make that kind of a mistake.

Troubleshooting a septic system is not a complicated process; there are only a few possibilities to consider. The two big situations you may encounter are a system that will not accept any more waste and a system that is flooded, as I described earlier.

It is helpful to have a competent understanding of septic systems when talking with property owners. For example, what would you say if a homeowner asked your opinion about installing a garbage disposer in a house served by a septic system? Some plumbing codes prevent such installations and others don't. What is the correct answer? Well, we'll discuss this question a little later, but there may not be a cut-and-dry answer available.

If a customer's property is being served by a septic system and you are asked to route storm-water drainage into the sanitary drainage system, would you do it? I hope not; I will expand on this in a moment.

Are bacteria additives necessary for properties connected to septic systems? If you don't know the answer to this question, you will soon.

Will chemicals flushed into a building's drainage system have any affect on the effectiveness of a septic system? They can, and I'll explain why shortly.

Could cold weather freeze a septic field and cause a backup in the sanitary drainage system? You bet it can, and I'll prove it in a few minutes.

Are you starting to see how issues related to septic systems can affect your daily plumbing career? Now that you've had a taste of the kind of questions you may be asked to answer as a plumber, let's take the questions, and a few others, and answer them on a case-by-case basis.

Should garbage disposers discharge into septic systems?

Many jurisdictions prohibit the installation of garbage disposers in properties served by septic systems. However, not all code enforcement offices take this stance. So what is the correct answer? I'm not sure anyone knows for sure, but let me give you some background information on the subject.

Garbage disposers can be responsible for putting heavy additional loads of solids into a septic system. If a disposer is used infrequently or only to handle small food particles, my guess is that no serious damage will occur to a septic system. If however, the disposer is required to digest large amounts of substantial food particles, it stands to reason that the extra load on a septic system may have some adverse effects.

I have seen written opinions that if a disposer is installed on a septic system, the size of the system should be increased by 50 percent.

I feel the biggest risk to using a garbage disposer with a septic system is the possibility of clogging the drainfield. If a drainfield becomes clogged, it normally has to be replaced, and that is an expensive proposition.

When I worked in Virginia, garbage disposers were prohibited on septic systems by the plumbing code. In Maine, disposers can be installed with septic systems, but I discourage the practice.

Should storm water discharge into septic systems?

Storm water should not be discharged into a septic system. If too much water is introduced into the septic tank, the tank can not break down its contents properly. Excessive water can also flood the drainfield, causing a stinking mess.

Will chemical discharge affect a septic system?

Chemical discharge can affect a septic system. Chemicals can inhibit the natural action of bacteria in a septic tank. It is unlikely that modest amounts of household chemicals would damage a septic system, but heavy concentrations could.

Are bacteria additives needed in a septic system?

I don't believe that bacteria additives are needed in a septic system, and there are many professionals who share my opinion. There may be some special circumstances when bacteria should be added to a septic tank, but the occasions, in my opinion, are rare, if they exist at all.

Do freezing temperatures affect a septic system?

Yes, freezing temperatures can affect a septic system. Bacteria in a septic tank is not as effective in cold weather. This causes the breakdown period of solids to be extended and can create some problems.

The biggest problem with freezing temperatures occur when the drainfield has been installed at a shallow depth. I've seen drainfields in Maine freeze to the point that they caused sewage to back up into homes. When this happens, the only reasonable thing for homeowners to do is to have the tank pumped frequently. This option is expensive if the field stays frozen for an extended time, but it is the only feasible solution during a long, cold, Maine winter.

How far should a drainfield be from a well?

Drainfields should be at least 100 feet from any potable water source, and they should be situated so that they will not drain to the water source.

Is it safe to flush paper towels into a septic tank?

It is not a good idea to flush paper towels into a septic tank. Paper towels, moist towelettes, and similar products may not be broken down in a septic tank. If these materials get into the drainfield, they can clog the field.

How can I find a septic tank to check it?

There are several clues that indicate the presence of a septic tank. If you are looking for the tank in warm months, look for lush, green grass. This indicates a probable location for the drainfield. Once you find the drainfield, look back toward the building for a depression in the ground.

When the tops of septic tanks are removed for cleaning they often leave a hollow in the ground after being replaced. A septic tank shouldn't be installed any closer than 8 to 10 feet from a building, but this rule does not always hold true, especially in the case of older tanks.

Checking to see where the building drain leaves the foundation of the building will give you a good idea of what direction to look in and how deep the tank is likely to be.

If you happen to own a metal detector, you can use it to find septic tanks. The rebar in the concrete that is used for handles on the lids will be easily picked up by a metal detector.

How can I tell if the septic tank is causing drains to back up?

You can tell if the septic tank is causing drains to back up by digging up the septic tank and removing the cover over the inlet. Have someone flush a toilet in the building while you watch for water at the inlet of the tank. If the water never gets to the tank, you can be sure the tank is not causing the problem. When the water level in the tank is above the inlet invert, you can bet the tank and/or the drainfield are responsible for the backed-up drains.

Is it true that there is gas in septic tanks?

There is usually gas collected in septic tanks. The gas is flammable and open flames should not be allowed near an uncovered tank. It is also advisable to avoid breathing the gas and fumes that rise from an uncovered septic tank.

In General

In general, plumbers will not be asked to correct problems with septic systems. It is advantageous, however, to understand how these systems work, and I hope you have a better appreciation for septic systems at this time. Let's move on now to Chap. 16 and study the troubleshooting methods required for water treatment systems.

16

Troubleshooting Water Treatment Systems

Water treatment systems are not new on the plumbing scene, but their presence is growing rapidly. As people become more aware of what is, or may be, in their water, the demand for water conditioning equipment expands. With the growth of this industry, plumbers must extend their knowledge of water treatment systems.

There are plenty of companies that specialize in water conditioning, but a lot of people depend on their regular plumber to handle the installation and service of water treatment equipment.

Not all plumbers are willing to work on filters, acid neutralizers, and general conditioning equipment. However, many plumbers see the profit potential offered from conditioning water, and they are eager to take on this type of work.

What do you know about water quality and water conditioning? If you don't have a broad knowledge of the subject, don't feel bad; many plumbers are weak when it comes to knowing the ins and outs of water treatment.

This chapter is going to help you understand water conditioning and the equipment used in doing it. To understand the mechanics of water treatment equipment, you should first know something about its purpose.

It is not hard to figure out that an acid neutralizer neutralizes acidic water, but what is acidic water? If water has a pH of 14, is it highly acidic? No, it isn't. A pH of 7 is neutral and a pH of 14 is alkaline. The lower the pH, the more acidic a substance is. For example, muriatic acid has a pH of 0 and vinegar has a pH of 3.

Acid is not the only reason water conditioning equipment is used. Iron and sulfur can cause trouble with plumbing and require water to

be conditioned. Hard water is a common problem that water treatment equipment can be used to solve. Let's look at the various reasons why water conditioning is necessary and explore the signs you,
as a troubleshooter, should look for.

Hard Water

Hard water is a common problem in many parts of the country. Hard
water is water that contains such elements as calcium and magnesium. When the content of these elements in water is high enough,
the water is considered hard and deserving of treatment.

Hardness is measured either by grains per gallon or by parts per
million. When the hardness of water is over 7 grains per gallon, a
water softener is often used to treat the hard water. The measurement of 7 grains per gallon is not a magic or conclusive number. Some
people don't worry about hard water until the count reaches 8 or 9
grains per gallon, and others will want their water treated if the
count is 5 grains per gallon. The determining factor of when to treat
hard water is usually measured by the results untreated water is
having on a plumbing system.

What are the trouble signs of hard water?

The trouble signs of hard water are fairly easy to spot. Homeowners
who live with hard water are forced to use more soap and water to
clean laundry than people who wash with softer water. The hard
water reduces the effectiveness of soap and can leave stains on clothing.

In addition to making laundry duties more difficult, hard water can
produce a film that covers dishes washed in the water. This same
type of film can coat plumbing fixtures such as like sinks and bathtubs. The dull, dirty residue left from hard water detracts from the
appearance of plumbing fixtures and makes dishes seem dirty even
after they have been washed.

Poor water pressure can be the result of hard water. When hard
water resides in pipes and water heaters, it leaves deposits that can
shorten the life of the plumbing system. Heating elements in electric
water heaters can become encrusted with these deposits, flush holes
in toilets can become clogged with the residue of hard water, and
pipes can gradually close up with the buildup left from hard water.

As a troubleshooter, you should remember these symptoms when
you are servicing or repairing a plumbing system. Now that you know
what to look for, you can correct the immediate problem caused by
hard water, and you can suggest a water test and possibly water
treatment equipment to solve the problem permanently.

Iron

Iron can cause a lot of trouble when too much of it is present in potable water. When iron is suspended in water, it creates a problem by staining objects it comes into contact with. The staining can occur on plumbing fixtures and clothes that are washed with the water.

Iron can be present in two different forms. It can be ferric oxide or ferrous oxide. As ferric oxide the iron content will produce rust stains and require treatment. When the iron appears as ferrous oxide, it is soluble, tasteless, colorless, and odorless.

Iron content is measured in parts per million. When the iron content is low, say under 0.3 parts per million, it is generally not a problem that requires treatment. However, as the content rises above 0.3 parts per million, the taste of the water may be objectionable and staining is likely.

When water is being derived from deep wells, its iron content may start out as ferrous oxide and change to ferric oxide. When the water is drawn into the plumbing system and exposed to air, it can change from its nice form to its nasty form. Carbon dioxide is capable of converting the ferrous oxide to ferric oxide.

Have you ever lifted the lid off the tank of a water closet tank and seen a black slime layer? The slippery slime is created by iron bacteria. This bacteria feeds on well pumps, pipes, and tanks that maintain a constant contact with the water holding the bacteria.

When the iron content of water exceeds acceptable levels, the water can be treated with a water softener. If the iron content is extremely high, an iron filter will be necessary.

Rotten Eggs

Have you ever run water in a sink and been surprised to smell the odor of rotten eggs? The smell is from water that contains a hydrogen sulfide solution, more commonly called sulfur. Water tainted with hydrogen sulfide will often take on a black color. It also tastes and smells like rotten eggs.

Sulfur water is hard to live with. It makes many drinks, like coffee and tea, taste bad. Mixing liquor with sulfur water can result in a black drink, and showering under the fumes of sulfur is no fun.

Aside from its unpleasant odor and taste, sulfur water is corrosive. The hydrogen sulfide works to destroy pipes, water tanks, water heaters, pump parts, and some plumbing fixtures. Anything made with an alloy of steel, iron, or copper can be adversely affected by sulfur water.

Hydrogen sulfide is measured in parts per million. Generally, water with a content of less than $\frac{1}{2}$ part per million does not require treatment. Water with a hydrogen sulfide content of 1 part per million is normally considered undesirable. Oxidizing filters are normally used to control hydrogen sulfide in water.

Acid

Acid can wreak havoc with a plumbing system. You saw earlier in the chapter how the pH rating of water is the measurement of its acidic content. The pH scale runs from 0 to 14. Water with a rating between 6 and 8 on the scale is considered best.

Typically, water will either be acidic or hard. Most water that contains less than 3.5 grains per gallon of hardness will be acidic, at least to some extent.

What does acidic water do to a plumbing system? It can eat away at the metal in the system. For example, acidic water can eat right through copper tubing. A sure sign that acidic water is present in copper water lines is the green stains that occur in plumbing fixtures. In older plumbing systems, where galvanized steel pipe was used for water distribution, the stain is rust colored. Acid neutralizers are used to control the acidic content in water.

Cloudy Water

Have you ever filled a glass with water and seen that the water was cloudy? Cloudy water is turbid. Turbidity includes muddy water, cloudy water, and water that contains sediment. Turbid water is common when water is drawn from a stream, but it is uncommon to be troubled with turbid water from a well.

Turbidity is measured in parts per million. Water with a rating of less than 10 parts per million is generally considered acceptable. In cases where water is too turbid, a sand filter can be used to correct the problem.

Now that we have discussed the most common reasons for treating water, let's look at the types of equipment used for the job.

Water Softeners

There are two basic ways to soften water. It can be done with an ion-exchange softening or a lime-soda treatment. We will start with the ion-exchange process.

Ion exchange

The ion-exchange procedure for softening water removes the calcium and magnesium that creates hard water. This is done through a chemical process.

Water that is being treated is passed through a bed of ion-exchange material, such as zeolite. The ion-exchange material traps and retains unwanted minerals in the water and replaces them with an equivalent amount of sodium.

When the ion-exchange material loses its ability to release sodium, it must be regenerated. This is done by introducing salt into the ion-exchange material. Once the salt is in place, the ion exchanger can once again produce sodium ions, and the system is back into satisfactory operating condition.

If you are troubleshooting an ion exchanger that is no longer softening water properly, investigate the ion-exchanger bed and regenerate it with salt if necessary.

Water softeners are rated in terms of grain capacity. A softener with a 20,000-grain capacity will require about 10 pounds of salt for regeneration.

How does a water softener work?

Most softeners go through four stages of operation: backwash, regeneration, rinse, and service. Backwashing usually lasts for about 10 minutes and is designed to loosen and dispose of deposits. Then the softener goes into a regeneration cycle; this is when new salt is added to the filtration bed. The rinse cycle is next, and it often runs for about an hour. Once the rinse cycle is complete, the water softener is back in service. When called upon, the softener leaves its service stage and goes into another backwash phase.

Lime-soda process

The lime-soda process of softening water is another way of controlling hard water. If lime or lime and soda ash are added to hard water, the hardness chemicals are forced to precipitate. They are then removed by sedimentation and filtration. The lime-soda process is rarely used for individual plumbing systems.

Iron and Manganese

Iron and manganese are common complaints for plumbing systems served by private water supplies. The same ion-exchange water softener described above will work to remove moderate amounts of iron.

If water has an iron content of more than 3 parts per million, an ion-exchange softener may not be able to condition the water satisfactorily. When the iron content is between 3 and 10 parts per million, a manganese zeolite filter should be able to control the iron problem. For water with higher concentrations of iron, it will be necessary to use an iron filter and/or a chlorinator.

A manganese zeolite filter removes both iron and manganese. This type of filter oxidizes the iron to precipitate it, and then filters it. Manganese dioxide in the filter is what does the job. Potassium permanganate is used to replenish the oxygen in the filter.

Iron that has been oxidized with chlorine can be removed with a sand or carbon filter. This type of treatment is good in that the chlorine also kills the iron bacteria that causes slime buildups. The bacteria cannot be removed by filtration alone.

It is also worth noting that iron does not oxidize well in acidic water. If you are dealing with acidic water and an iron problem, neutralize the pH of the water before chlorinating the iron.

Removing Sulfur

Removing sulfur in water can be done with a manganese zeolite filter or with chlorination. A manganese zeolite filter can be use for water with a sulfur content of up to 5 parts per million. The filter oxidizes the sulfur, turning it tasteless, odorless, colorless, and noncorrosive.

Chlorine will also oxidize sulfur, and it is probably used more often than manganese zeolite filters. The downside to chlorinating is that so much chlorine is required to get the job done that the chlorinated water must be filtered through a carbon filter to remove excess chlorine. The installation of the filter is not extremely costly or troublesome, but it is something you must be aware of.

Overcoming Acidic Water

There are two common ways to overcome acidic water. Acid neutralizing filters are quite common. When acidic water is run through these filters, the pH value is raised. This is done with a chemical reaction between calcium carbonate and carbon dioxide. However, acid neutralizers can create a new problem as they are solving the existing problem of acidic water; they can create hard water. The calcium carbonate used to beat the acid dissolves in the water and increases the hardness rating.

Soda-ash solutions are another way to neutralize acidic water. This process does not increase the hardness of water. The soda ash combines chemically with water to neutralize it.

Sand Filters

Sand filters are very effective in treating turbid water. There are times, however, when the sediment particles causing turbidity are too small for a filter to catch. For an inexperienced troubleshooter, this situation can be very frustrating.

Chemicals can be added to water to enlarge the size of suspended particles so that they may be trapped by a filter. Aluminum sulfate is one such chemical.

Tastes and Odors

Unpleasant tastes and odors are usually controlled with carbon filters. The activated carbon in these filters has been treated to increase its power of absorption. When the carbon can no longer absorb impurities, it must be replaced.

Equipment Specifications

Equipment specifications can vary from one manufacturer to another. In this section of the chapter we are going to look at some specific equipment statistics. We will start with a combination water softener and iron filter.

The water conditioner we are about to discuss is one of the models my company offers to customers. Let me tell you a little about what this piece of equipment is capable of, and then I will go into the details of what makes the unit tick.

High-capacity water softener and iron remover

The first unit we are going to take a look at is a high-capacity water softener and iron remover. This combination water softener and iron filter can remove hardness and most of the dissolved iron in a water supply. It can also filter suspended iron and may or may not remove organically bound iron. If the iron content is high, above 2 parts per million, the mineral bed may need to be cleaned periodically.

This cleaning can be done with a mineral bed cleaner or by mechanical cleaning. Once the cleaning is completed, the unit is back to its original capacity.

The unit has no influent restriction on the pH level and will perform well throughout the 0-to-14 range of the pH scale. Free chlorine is limited to and should not exceed 1.0 part per million.

On this particular model, there is an adjustable head. The brass control valve is a vital part of this system. This valve is fully adjustable for cycle times and regeneration settings, which is a big advantage when the raw water conditions change and require adjustments in the head. This one factor makes this unit better than many average systems. A unit that is not equipped with an adjustable head requires extensive work if existing raw water conditions change. By having a unit that has an adjustable head, the corrections for changing water conditions can be compensated for.

Let's do an inventory of the system components for this type of unit. The system consists of a softener tank and a brine tank. The softener tank is made up of a control valve, a 1-inch distributer tube, a mineral tank, and a high-capacity resin.

The brine tank has a brine well cap, a 4-inch brine well, a brine valve, a brine air check, a red and white brine float-rod, a brine overflow fitting, a salt support plate, and a 4-inch leg set.

Different models of this unit have different ratings, but I will give you some technical data on the mid-range model. This unit has an ionic-exchange capacity of 45,000 grains. It uses 12 pounds of salt to recharge itself. The unit has $1\frac{1}{2}$ cubic feet of mineral and a backwash flow rate of 2.4 grains per minute. The brine flow rate is 0.5 grains per minute. With a $\frac{3}{4}$-inch inlet, the flow rate at its peak is 9 grains per minute.

Standard-capacity water softener and iron remover

The second unit we are going to examine is a standard-capacity water softener and iron remover. This unit can provide efficient care of water that contains hardness, iron, manganese, hydrogen sulfide, or acid.

The mineral zeolite is used in this model to solve water problems. This piece of equipment can remove 17,000 grains of hardness per cubic foot, 10 milligrams of iron and manganese combined, and 1 milligram of hydrogen sulfide gas and can raise the pH by sacrificing silica.

High contents of iron can cause problems. In the case of an installation with high iron, a resin bed cleaner should not be used because it can deplete the mineral bed quickly. A concentration of regular household bleach will do the job without the same risk of premature depletion of the mineral bed.

This model, like the one previously described, relies on an adjustable control head to maintain an ongoing treatment of water that is satisfactory.

Automatic iron filter

The next piece of equipment we will take a close look at is an automatic iron filter. This unit is fully automatic, and it is used to oxidize and filter dissolved and suspended iron from residential water.

A formulated glauconite greensand is used to remove high concentrations of iron, manganese, and hydrogen sulfide. A weak solution of potassium permanganate is rinsed through the unit periodically to restore the oxidizing ability of the filter.

Influent water should have a pH level of at least 6.5 to 8.8 and an alkalinity of at least 4 milligrams for each milligram of iron or manganese.

Sanitizing with chlorine is not a problem when this type of equipment is used. A chlorine feed will enhance the precipitation of the iron and lower the consumption of the potassium permanganate.

An adjustable control head, like those previously mentioned, makes this unit very versatile in its ability to adapt to changing water conditions.

The two parts of this system are a filter tank and a pot perm tank. The filter tank is equipped with a control valve, a 1-inch distributer tube, a mineral tank, manganese greensand, and a bottom gravel support bed.

The pot perm tank has a brine well cap, a 4-inch brine well, a pot perm valve, an overflow fitting, and a felt pot perm pad.

In a mid-range size, this unit has an oxidation capacity of 15,000 milligrams. The pot perm tank uses 4.5 ounces of material for each regeneration and a mineral amount of 1.5 cubic feet. The backwash flow rate is 5 grains per minute, and the inlet flow rate, through a $\frac{3}{4}$-inch inlet, is 9 grains per minute.

Acid neutralizers

The acid neutralizers I recommend are fully automatic and can be used on private water sources and on municipal water. The units depend on a crushed and screened neutralizing media called neutralite. When acidic water comes into contact with the neutralite, it dissolves the calcium carbonate media slowly to increase the pH level.

The mineral level in this acid neutralizer becomes depleted as the unit is used and backwash cycles continue. At some point, depending upon the use and severity of the water conditions, the neutralite must be replenished.

A common side effect of an acid neutralizer is that the water being neutralized becomes harder. Normally, this increase in hardness is not enough to cause problems. If, however, the hardness reaches a point of concern, a water softener should be installed with the acid neutralizer to condition the water fully.

In its standard form, this neutralizer is effective on water with a pH rating as low as 6 on the pH scale. For more acidic water, a blend of Corosex is added to compensate for the increased acidity.

Again, the adjustable control head found on this unit is its most outstanding feature. It allows complete control over the settings for cycle times and regeneration settings.

In terms of construction, the acid neutralizer consists of a control valve, a 1-inch distributer tube, a mineral tank, a neutralite, and a bottom gravel support bed.

The technical ratings on a mid-range unit are as follows: 1.5 cubic feet of mineral amount, 5 grains per minute in the backwash cycle, and 7 grains per minute at the peak flow of the inlet through a ¾-inch pipe.

Carbon filters

Carbon filters are next on our list of equipment to study. These units are used to remove tastes, odors, and colors from water. The fully automatic carbon filters that I recommend use a granulated activated carbon which has a high density. Because of the make up of the pore structure, the unit is balanced and provides the ability to offer an efficient absorption of many odors and tastes.

People who suffer from water tainted with the smell of rotten eggs need carbon filters. This condition also frequently calls for a chlorine cleaning. The chlorine can present its own problems with odor, but a good carbon filter is capable of eliminating the smell of chlorine. When a good unit is installed, the free chlorine will be reduced to clean, odor-free water.

Backwashing intervals regrade the media and keep it clean of accumulated suspended matter. This cleansing process also prevents losses in water pressure that can be caused by clogged filters.

While activated carbon filters have a high capacity for taste and odor removal, the media in the filters needs to be replaced from time to time. The length of time between replacements is generally determined by the severity of the water conditions being treated.

As with all the other water conditioning units I recommend, this automatic carbon filter is equipped with an adjustable control head.

The key parts of a carbon filter include the control valve, a 1-inch distributer tube, a mineral tank, the activated carbon, and a bottom gravel support bed.

The technical stats on this carbon filter show that a mid-range system has a flow rate of 5 grains per minute in backwash, a 7-grains per minute flow rate from a ¾-inch inlet, and 1.5 cubic feet of mineral amount.

Sediment filters

The type of sediment filters that I like are fully automatic and use a nonhydrous aluminum silicate media. This gives the unit a super ability to remove high concentrations of dirt from water. The irregular surface characteristics in the media of the model I favor is what makes it so effective.

As added benefits, this type of filter offers lower pressure drops, less backwash capacity, higher service flow rates, and longer service runs than most conventional media, such as quartz sand. With a bed depth of 30 inches, this unit is capable of great filtration service.

I know it's beginning to seem redundant, but this piece of equipment also comes with a control head that is fully adjustable. These heads cost a little more than their preset, fixed competitors, but they are worth every penny to the plumber in the field who must make adjustments for changing water conditions.

In terms of construction, the sediment filter consists of a control valve, a 1-inch distributer tube, a mineral tank, a filter, and a bottom gravel support bed.

The statistics on a mid-size unit show a 1.5-cubic-foot mineral amount, a backwash flow rate of 5 grains per minute, and an incoming flow rate of 7 grains per minute through a ¾-inch inlet.

Birm filters

Birm filters are used to remove dissolved iron and manganese. The media used in my favorite birm filter is a granular filter product known simply as Birm. It acts as an insoluble catalyst to aid in the reaction of dissolved oxygen with the dissolved iron and manganese to create hydroxides. The hydroxides precipitate out and are then mechanically trapped and removed through routine backwashing. The filter system is fully automatic.

Chemicals are rarely needed with this type of system. The periodic backwashing is all that is normally needed. However, the influent water for treating iron must not contain oil or hydrogen sulfide. The dissolved oxygen content should be 15 percent of the total iron content, and the pH should be rated at a number of at least 6.6.

When treating manganese, the oxygen content should still be at 15 percent of the total iron and manganese. The pH rating should fall somewhere between a figure of 8 and 9. If the pH must be raised, soda ash is a suitable substance for getting the job done.

Like all my other pet water conditioning equipment, this unit is also equipped with an adjustable control head. In addition to the head, there is a 1-inch distributer tube, a mineral tank, regular birm, and a bottom gravel support bed.

The specifications on this model, in a mid-range size, call for 1.5 cubic feet of mineral amount, 5 grains per minute on the backwash cycle, and 7 grains per minute on the incoming water from a ¾-inch inlet.

Multimedia

Multimedia filter systems are needed when you are faced with difficult water conditions. If your customer insists on a system that doesn't use chemicals, a multimedia filter is the place to turn for help.

Since multimedia filters don't depend on chemicals to treat water, they rely on a number of different types of media. A common multimedia filter will contain Birm, neutralite, filter-ag, Corosex, and an aeration package. These components are put together in a special combination.

Birm in the system acts as a catalyst with iron and dissolved oxygen to cause oxidation and precipitation. Neutralite increases the pH rating of water and corrects acidic conditions. Filter-ag is used to trap suspended matter, and the addition of Corosex elevates pH levels to encourage the removal of manganese.

The system is equipped with an aeration package that provides oxygen to the water. This increased oxygen level assures that all the components will work together to perform their process of oxidation, precipitation, and filtration.

This type of nonchemical system relies heavily on backwashing, which cleans the filter bed of precipitates. The sizing of such a system depends on the available backwash water capability.

Influent water should have a pH rating in the range of 6 to 8.5. Iron bacteria should not be present in the influent water, and the sulfur concentration should be around 5.0 milligrams, depending upon the pH level of the water.

As you might imagine, the system I prefer is equipped with an adjustable control valve. There is a 1-inch distributer tube, a mineral tank, the multimedia, and a gravel support bed. The aeration package consists of an air release valve, an aerator, and the aeration tank.

In terms of specifics, you can expect a backwash flow rate of 5 grains per minute, a mineral amount of 1.5 cubic feet, and an inlet flow, through a ¾-inch inlet, of 7 grains per minute. These numbers are based on a mid-size system.

Combination units

Combination units can perform multiple functions. One such unit works to guard against iron-tasting water, sulfur smells, rust stains, and corrosion. Examine Figs. 16.1 through 16.4 to see how this device does its job. Also, keep in mind the value of adjustable controls on water conditioning equipment.

Reverse osmosis

Reverse osmosis systems are used to reduce dissolved solids in potable water. These compact units can be mounted under a standard sink.

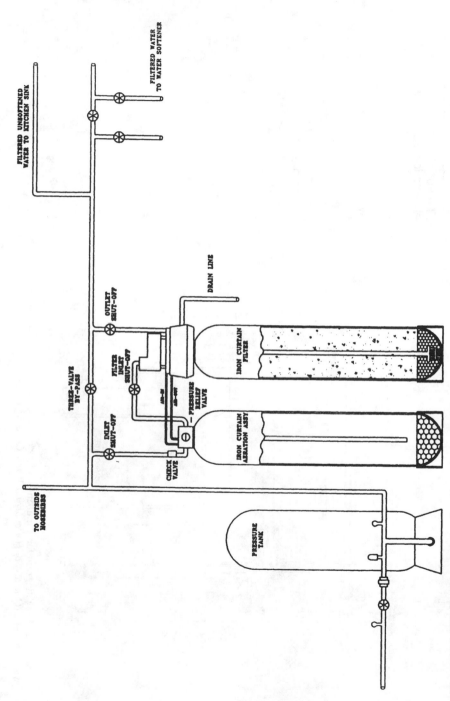

Figure 16.1 Piping diagram for the "Iron Curtain." (*Courtesy of Helienbrand Water Conditioners, Inc.*)

329

Step 2. Backwashing the Multi-Media Filter Bed

The Iron Curtain Control Center automatically shifts the controller into the backwashing cycle. Raw water enters the filter valve inlet from the aeration system, flows down thru the distribution system and up thru the multi-media bed, and out the drain line. The oxidized contaminents which were removed during the service cycle are backwashed out to the drain.

Patent Pending

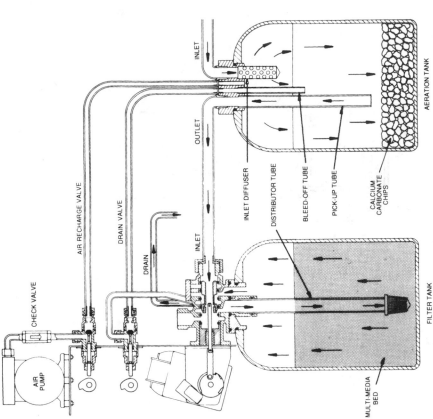

CHECK VALVE

AIR PUMP

AIR RECHARGE VALVE

DRAIN VALVE

DRAIN

INLET

INLET

OUTLET

INLET DIFFUSER

DISTRIBUTOR TUBE

BLEED-OFF TUBE

PICK-UP TUBE

CALCIUM CARBONATE CHIPS

AERATION TANK

MULTI-MEDIA BED

FILTER TANK

Figure 16.2 Flow diagram for a multiuse water conditioner (the Iron Curtain) in the backwash position. (*Courtesy of Hellenbrand Water Conditioners, Inc.*)

Step 4. Recharge of Aeration Tank with New Air

After approximately 2-4 minutes a specially mounted switch in the Iron Curtain Control Center turns the air compressor pump on for approximately 10 minutes. Water and/or air continue to run to the drain thru the drain valve and the compressor pumps air thru the air recharge valve, directly into the aeration tank. At the end of this cycle the aeration tank has been recharged with fresh air, and will have approximately an 18 inch head of air.

Patent Pending

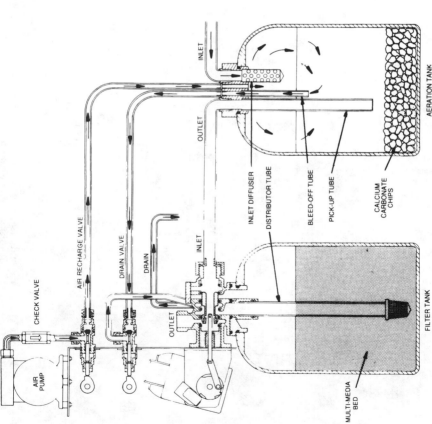

CHECK VALVE

AIR RECHARGE VALVE

DRAIN VALVE

DRAIN

AIR PUMP

OUTLET

INLET

INLET

OUTLET

INLET DIFFUSER

DISTRIBUTOR TUBE

BLEED-OFF TUBE

PICK-UP TUBE

CALCIUM CARBONATE CHIPS

AERATION TANK

MULTI-MEDIA BED

FILTER TANK

Figure 16.3 Flow diagram for a multiuse water conditioner (the Iron Curtain) in the recharge position. (*Courtesy of Hellenbrand Water Conditioners, Inc.*)

Step 6. Packing the Bed for Filtering

The Iron Curtain Control Center automatically closes both the drain and air recharge valves and shifts the piston into the rapid rinse position. Raw water passes thru the aeration system and enters the control valve, passes down thru the multi-media filter bed into the bottom distribution system, up thru the distributor tube, and out to the drain. At the end of this cycle the Iron Curtain System automatically returns to the service position.

Patent Pending

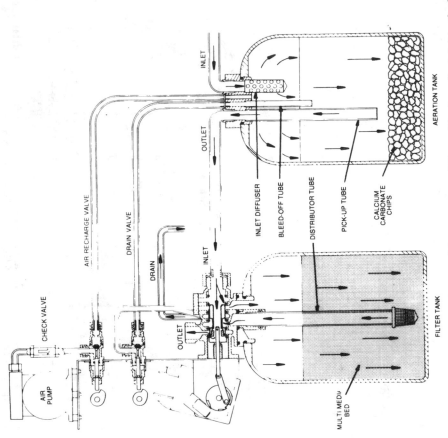

CHECK VALVE

AIR PUMP

AIR RECHARGE VALVE

DRAIN VALVE

DRAIN

OUTLET

INLET

INLET

OUTLET

INLET DIFFUSER

BLEED-OFF TUBE

DISTRIBUTOR TUBE

PICK-UP TUBE

CALCIUM CARBONATE CHIPS

AERATION TANK

MULTI MEDIA BED

FILTER TANK

Figure 16.4 Flow diagram for a multiuse water conditioner (the Iron Curtain) in the bed-packing position. (*Courtesy of Hellenbrand Water Conditioners, Inc.*)

Changing Water Conditions

Changing water conditions can wreak havoc with existing water conditioning systems. Many systems are not equipped with adjustable control heads. These systems are preset for specific duties, and they cannot be fine-tuned to react to periodic changes in water conditions. This can create some real problems for plumbers who are not experienced in water treatment.

When water levels fluctuate, and they do so frequently, different water conditions can occur. A home that had a tolerable level of hydrogen sulfide before the changes in the water table may take on odors or taste that can no longer be tolerated.

I have just been involved in such a situation. The home that I was called to was built many years ago, and it has never been equipped with water treatment equipment. The house is located in a small subdivision, where the neighbors are close by.

The neighbor to the side of the house I've been working in has no problems with the quality of water being delivered from his well. The house across the street from my customer has a significant problem with the smell of rotten eggs. Those people, according to my customer, buy bottled water for cooking and drinking and have learned to live with the odor when bathing in their household water.

My customer loves to drink water, and she hates the effect the hydrogen sulfide is having on her. She has had minor problems with the sulfur in the past, but chlorine treatments in her well on an annual basis have controlled the problem in the past. Now, however, the chlorine treatments are not lasting long, and the smell of rotten eggs permeates her home.

Being frustrated by her problem, the homeowner sought professional help. The first people she talked to recommended a simple, in-line filter to solve her problem. When I inspected her house recently, I found that the filter the other so-called professionals had sold her was a simple sediment filter.

The odor in her water is so bad that with water running from her kitchen sink, the smell drifts throughout most of the home. We're talking serious sulfur problems here.

Before calling me, the lady called some other plumbers and water treatment companies. The plumbers did not impress her with their knowledge of water treatment, so she homed in on the water conditioning company. The company is one that enjoys the benefit of a major franchise name.

I'm not sure what went wrong between the homeowner and the water treatment company, but she wound up feeling that she might be being taken advantage of. It was then that she called my company.

I responded to her call to do a free water analysis. The results of her test have not come back yet, but I'm sure she is suffering from both a hydrogen sulfide and iron content problem. There is no way a simple $25 hardware-store sediment filter is going to handle such a problem.

This is a prime example of how changing water tables can affect existing water treatment equipment. Even though this particular customer does not have professional water conditioning equipment installed, the changing water tables we have experienced in Maine this summer have forced her into action. Her problem used to be tolerable with a few chlorine treatments, but now it has escalated to a point where professional treatment equipment is needed. This same type of problem can disturb water treatment systems that worked well in past years.

The types of conditioning equipment I have described in detail, with adjustable control heads, allow professionals to monitor water conditions periodically and adjust the conditioning system to compensate for changes in the water.

Adjustable control heads are a little more expensive than preset controls, but if the water conditions ever change substantially, the small additional cost for the adjustable head more than pays for itself in both convenience, service, and reduced labor charges for the customer.

A Golden Opportunity

I believe a golden opportunity awaits those who become established in the field of water treatment and conditioning. It is unfortunate, but it is likely that our potable water supplies will continue to get worse as time goes on. This fact combined with the education of consumers will increase the demand for water treatment, both on private water supplies and on municipal water services. The niche available in water conditioning can be a wonderful expansion opportunity for any plumbing company.

Getting started in water treatment services is not expensive or complicated. You don't have to buy a major franchise to cash in on the big money being invested in water conditioning equipment. If you are not comfortable with your working knowledge of water conditioning equipment, you may be able to attend a class or seminar that will bring you up to speed.

In my area, local suppliers offer free training and troubleshooting seminars on the subject of water treatment. I have attended some of these instructional meetings, and I've found them to be both enlightening and thorough.

If you choose a good supplier for your equipment, you will enjoy the benefit of having experts just a phone call away. For example, the company I deal with has a resident engineer available to answer any questions that come up and to help in the design of a solid working system.

All I have to do is take a water sample and send it to the company. They have the water tested and send out a letter recommending the best course of action. If the water sample shows cause for treatment, the company explains in their letter exactly what model and type of equipment should be installed.

I sit down with the customers and go over the test results with them. The customers are shown the letter that indicates the test results and equipment recommendations. If the customers decide to have the work done, I call the supplier and the equipment is delivered promptly. Then it is just a matter of installing the units.

If a problem arises in the field that one of my plumbers can't figure out and I don't have the answer, a phone call to the supplier solves my problem. If necessary, they will send out a company representative to troubleshoot the system.

I pay a little more for my equipment than some of my competition does, but I do so for two reasons. The equipment I use is top-notch, and the support the supplier offers is invaluable.

As you can see, a plumber doesn't have to know a tremendous amount about water conditioning if a supplier like mine is available. I don't believe you should venture into water conditioning blindly if you are not familiar with the equipment and the ways in which it works. However, neither do I think that you should ignore the lucrative earnings possible from water treatment because you feel intimidated by a lack of knowledge.

Getting into water treatment

Getting into water treatment is not difficult. If you run some advertisements or do some direct mailing that offers a free water analysis, you should get a good response. Since most suppliers are more than happy to provide test results without charge, your only expense is the advertising and the time it takes to collect the water sample.

Some plumbers feel this type of offer is too expensive to be worthwhile. They can't understand why anyone would go out to jobs and collect samples without charging for it. There are two good reasons for offering free water tests.

When you go out to collect a water sample, you are making contact with new potential customers. Even if they don't need or don't buy a water conditioning system from you, there is a good chance you will have gained new service customers. When something goes wrong with those customers' plumbing, the odds are good you will be called to handle the repair. You will be remembered for your free water test and the impression you made on the customer when you collected the water.

A percentage of all free water tests will result in the sale of conditioning equipment. The profit from these jobs should more than offset the time lost on water tests that do not provide work.

Now that you've had my sales pitch on the benefits of water treatment services, let's get back on track with troubleshooting. If you will turn to the next chapter, we will investigate the many problems you may face when working with a water distribution system.

Troubleshooting Water Distribution Systems

Troubleshooting water distribution systems involves many problem possibilities. These systems can be simple residential installations or complex commercial jobs. Since a water distribution system consists primarily of pipe and fittings, many plumbers take the troubleshooting process for these piping layouts for granted.

It would seem logical that finding trouble with a water distribution system wouldn't be difficult, but it can be. If you've been in the trade for a number of years, you know I'm speaking the truth. Plumbers new to the trade may doubt the level of difficulty in troubleshooting water distribution systems, but they shouldn't.

I've been in the trade about 20 years, and I've spent many hours looking for problems in water distribution systems. Sometimes I've found them quickly, and sometimes I've been totally perplexed. Over all of these years I've wasted a lot of time because I didn't know where to look for the problems. With the experience I've gained, the hard way, I've learned where to look for plumbing problems and what possibilities to rule out quickly. This saves me time and my customers money.

With the help of this chapter, you will learn what to look for, where to look for it, and when to look for it when troubleshooting water distribution systems. As we move through this chapter, we will have to cover a lot of ground. There are a myriad of possibilities for problems with water distribution systems, and we will cover most, if not all, of them in the following pages.

Are you wondering what types of problems might crop up with a water distribution system? Well, I don't want to give away all of the secrets so soon, but I will give you a few hints of what we will be looking for.

Have you ever had a problem with copper tubing that developed pinhole leaks for no apparent reason? When was the last time a customer asked you to solve the problem of banging water pipes? Has water pressure ever been a complaint you have had to deal with? Do you know what to do when a house has frequent problems with faucets that leak? Well, we are going to answer all of these questions and a lot more as we travel through this chapter.

Noisy Pipes

Noisy pipes can make living around a plumbing system very annoying. Sometimes the pipes bang, sometimes they squeak, and sometimes they chatter. This condition can be so severe that living with the noise is almost unbearable. What causes noisy pipes? Water hammer is one reason pipes play their unfavorable tunes, but it is not the only reason.

Not all pipe noises are the same. The type of noise you hear gives a strong indication to the type of action that will be required to solve the problem. You must listen closely to the sounds being made in order to diagnose the problems properly.

What will you be listening for? The tone and type of noise being made will be all that you may have to go on in solving the problem of a water hammer. Let's look at the various noises individually and see what they mean and how you can correct the problem.

Water hammer

Water hammer is the most common cause of noisy pipes in a water distribution system. Banging pipes are a sure sign of water hammer. Can water hammer be stopped? Yes; there are ways to eliminate the actions and effects of water hammer, but implementing the procedure is not always easy.

What causes water hammer? Water hammer occurs most often with quick-closing valves, like ballcocks and washing-machine valves, but it can be a problem with other fixtures. The condition can be worsened when the water distribution pipes are installed with long, straight runs. When the water is shut off quickly, it bangs into the fittings at the end of the pipe run or at the fixture and produce the hammering or banging noise. The shock wave can produce some loud noises.

If a plumbing system is under higher than average pressure, it can be more likely to suffer from water hammer. There are several ways to approach the problem to eliminate the banging. To illustrate these options, I would like to put the problems into the form of real-world situations. I plan to share my past experiences with you in an entertaining and informative way.

A troublesome toilet

This first example of a water hammer problem involve a troublesome toilet. The residents of the home where the toilet was located hated the banging noise their water pipes made on most occasions when the toilet was flushed.

The toilet was located in the second-floor bathroom, and when it was flushed, it would rattle the pipes all through the home. There came a day when the homeowners decided they could not live with the problem any longer. They were fed up with the annoying banging of the pipes every time the toilet was flushed.

The couple called in a plumber to troubleshoot their problem. The plumber looked over the situation and went about his work in trying to locate the cause of the problem. He knew they were suffering from a water hammer situation, but he wasn't sure how to handle the call. In fact, the plumber gave up, washed his hands of the job, and left.

Distraught, the homeowners called my company. One of my plumbers responded to the call and quickly assessed that the upstairs toilet was the culprit. He recommended that the couple allow him to install an air chamber in the wall, near the toilet.

At first, the couple was reluctant to give their permission to my plumber to cut into their bathroom wall. Under the circumstances, however, there were not many viable options, so the homeowners gave their consent.

My plumber made a modest cut in the bathroom wall, around and above the closet supply, and installed an air chamber. This particular homeowner was one who wanted to know every move that was being made, and he was not too sure the plumber was doing anything that would really improve the situation.

After the air chamber was installed, my plumber activated the system and began to flush the toilet. After several sequences of flushing the pipes remained quiet; there was no banging. The homeowner had been doubtful of my plumber's decision, but he did admit the work solved the problem, and he was very happy.

Air chambers are frequently all that are needed to reduce or eliminate banging pipes (Fig. 17.1). My plumber was correct in his diagnosis of the problem and in effecting an efficient cure for the problem. It is possible, however, that the problem could have been solved by installing a master unit for a water hammer arrester in the basement of the home, eliminating the need to cut into the bathroom wall. There is, however, no guarantee that a master unit would have controlled the situation on the second floor.

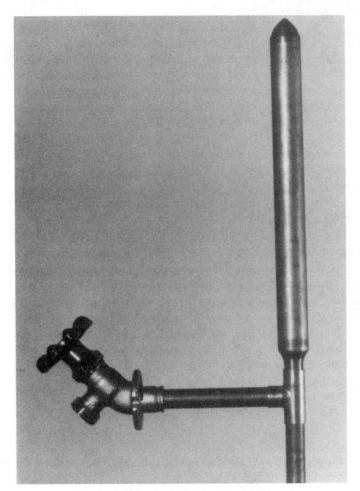

Figure 17.1 An air chamber installed for a hose bibb.

The plumber made a wise and prudent decision. By installing the air chamber at the fixture, satisfaction was practically guaranteed. While the master unit may have worked, there was some risk that it wouldn't. If the homeowner had objected to opening the bathroom wall, I'm sure the plumber would have installed a master unit in an attempt to control the problem.

Existing air chambers

Sometimes existing air chambers become waterlogged and fail to function properly. This problem is not uncommon, but it is a bother. If you have a plumbing system equipped with air chambers that is being affected by water hammer, you must recharge the exiting air chambers with a fresh supply of air. This is a simple process.

To recharge air chambers, you must first drain the water pipes of most of their reserve water. This can be done by opening a faucet or valve that is at the low end of the system and one that is at the upper end.

Once the water has drained out of the water distribution pipes you can close the faucets or valves and turn the water supply back on. As the new water fills the system, air will be replenished in the air chambers. As easy as this procedure is, it is all that is required to recharge waterlogged air chambers.

Unsecured pipes

Unsecured pipes are prime targets when you are having a noise problem. If the pipes that make up the water distribution system are not secured properly, they are likely to vibrate and make all sorts of noise. Not only is this annoying to the ears, it can be damaging to the pipes. Pipes that are not secure in their hangers can vibrate to the point that they wear holes in themselves. If enough stress is present, the connection joints may even be broken.

Local plumbing codes dictate how far apart the hangers for pipe may be. If the hangers fall within the guidelines of the local code and secure the pipes tightly, no problems should exist. However, when the hangers are farther apart than they should be, are the wrong size, or are not attached firmly, trouble can develop.

Pipes that are not secured in the manner described by the plumbing code may create loud banging sounds. This noise imitates the noise made by a system that is experiencing a water hammer. Squeaking and chattering noises may also be present when pipes are not secured properly.

The problem of poorly secured pipes is easy to fix, if you have access to the pipes. Unfortunately, the pipes and their hangers are often concealed by finished walls. Sometimes the problem pipes can be accessed in basements or crawlspaces, but as often as not, they are inaccessible, unless walls and ceilings are destroyed.

Once you have access to the pipes that are not secured in the proper fashion, all you have to do is add hangers or tighten the existing hangers. This is fine if you can get to the pipes easily, but it is a hard-sell to a homeowner who doesn't want the walls and ceilings of a home cut open. There is, however, no other way to eradicate the problem.

That squeaking noise

That squeaking noise you often hear from water pipes is almost always a hot-water pipe. Whether the water distribution pipe is copper or plastic, it tends to expand when it gets hot. As hot water moves through the pipe, the pipe expands and creates friction against the pipe hanger.

The expansion aspect of hot-water pipes is not practical to eliminate, but you can do something about the squeaking noise. The simple solution, when you have access to the pipe and hanger, is to install an insulator between the pipe and the support. This can be a piece of foam, a piece of rubber, or any other suitable insulator. Your only goal is to prevent the pipe from rubbing against the hanger when expansion in the pipe occurs.

Chattering

The chattering heard in plumbing pipes is not caused by cold temperatures; it is caused by problems in the faucets of fixtures. While this is more of a faucet and valve problem than a water distribution problem, we will cover it here since the noise is often associated with water pipes.

When you hear a chattering sound in a plumbing system, you should inspect the faucet stems and washers at nearby faucets. You will probably find that the faucet washer has become loose and is being vibrated, or fluttered as some plumbers say, by the water pressure between it and the faucet seat.

To solve the problem of chattering pipes, all you should have to do is tighten the washers in the faucets. Once the washers are tight, the noise should disappear.

Muffling the system at the water service

When you have a plumbing system that is banging because of water hammer, you can try muffling the system at the water service. This procedure doesn't always work, but sometimes it does.

Access to piping in many buildings is not readily available. Under these conditions the only easy way to attack a water hammer is at the water service or main water distribution pipe. By installing air chambers or water hammer arresters on sections of the available pipe, you may be able to eliminate the symptoms of water hammer throughout the building.

To install an air chamber or water arrester on the water service or main water distribution pipe is not a big job. Once the water is cut off, you simply have to cut in some tee fittings and install the air chambers or arresters. The risers that accept these devices should be at least 2 feet high, when possible.

Depending upon the severity of the water hammer problem, several devices may need to be installed at different locations along the water piping. Typically, one at the beginning of the piping, one at the end of the piping, and devices installed at the ends of branches will greatly reduce, if not eliminate, the problems associated with water hammers.

Adding offsets

Adding offsets to long runs of straight piping is another way to reduce the effects of water hammer. Long runs of piping invite the slamming and banging noise of a water hammer. If you cut out the straight sections and rework them with some offsets, you increase your odds of beating the banging noises.

Pinhole Leaks in Copper Tubing

Pinhole leaks in copper tubing are not uncommon in rural areas. These leaks can occur anywhere within the water distribution system, and they can cause a lot of damage to building materials, floors, and other items located in the buildings. These small leaks may drip a little water, or they may be large enough to produce a steady spray.

What causes pinhole leaks in copper tubing and pipe? Acidic water is the major cause of such problems. Many plumbing systems that derive their water supply from wells suffer from some degree of acidic water. When the pH rating of the water is low enough, the water can eat holes in the copper water pipes. It can also work to destroy faucet stems, pump parts, water tanks, fittings, and other elements of the plumbing system.

When a thin-wall copper tubing, like type M copper, is used to carry acidic water, it can become very tender. Squeezing the pipe with a pair of pliers can result in a crushed pipe and quite a leak. There are usually signs of acidic water that show up before the copper tubing is damaged. The evidence of acid being in the water is usually in the form of a blue-green stain in the plumbing fixtures.

Acid in the water distribution system will eat away at the bibb screws in faucets and other metal in the plumbing system. The deteriorating action of the acid can, in time, literally eat a plumbing system up from the inside out.

As you learned in Chap. 16, an acid neutralizer will control the acid in a potable water supply and eliminate pinhole leaks and diminished plumbing parts.

Low Water Pressure

Low water pressure often becomes a problem in some plumbing systems. Trying to take a shower in a home with low water pressure is very frustrating, and waiting several minutes for a water closet to refill its tank can be nerve-racking. There are many reasons, to be sure, why low water pressure will induce customers to call in professional plumbers.

What causes low water pressure? Ah, that's a loaded question; there are many possible causes for low water pressure. For buildings using pump systems, it could be a problem with the pressure tank or the pressure switch. Buildings on city water supplies may have pressure-reducing valves that need to be adjusted. Properties that are plumbed with old galvanized steel piping could be suffering from rust obstructions in the pipes. There are many reasons why a plumbing system may not have the desired water pressure. To expand on this, let's look at most of the reasons under a microscope.

Pressure tanks

Properties that obtain their water from private water sources depend on pressure tanks to give them the working pressure they want from their plumbing systems. If the pressure tanks become waterlogged, they cannot produce the type of water pressure they are designed to provide.

A waterlogged pressure tank gives a symptom that is hard for a serious troubleshooter to miss. When a pressure tank is waterlogged, the well pump cuts on very frequently, often every time a faucet is opened.

Sometimes waterlogged pressure tanks provide adequate pressure, even though they are forcing the pump to work much harder than it is intended to. At other times, there is a noticeable loss in pressure.

If you are faced with a building that has unsatisfactory water pressure and a well pump that cuts on frequently, you should take a close look at the pressure tank.

When a pressure tank is suspected of being waterlogged, it should be drained, recharged with air pressure, and then refilled with water. This is a simple process, and it can solve your pressure problems.

Pressure switches

Properties that are served by private wells depend not only on pressure tanks for their water pressure, but on pressure switches as well. If the cutin pressure on a well system is not set properly, it is possible for pressure to drop to unacceptable levels before the pump will produce more water.

A fast visual inspection of the pressure gauge on the well system will tell you if the system is maintaining a satisfactory working pressure. It is important, however, to make sure there is demand from a plumbing fixture while you are observing the pressure gauge.

If no plumbing fixtures are being called upon for water, the pressure gauge will remain static at its highest level. To be sure the system is maintaining an acceptable working pressure, you must put a demand on the system.

A pressure gauge that shows a sharp fall in pressure before the pump cuts on indicates that the cutin pressure is set too low. Going into the box of the pressure switch and adjusting the spring-loaded nut will correct this problem.

Pressure-reducing valves

Pressure-reducing valves that are not set properly can cause trouble in the form of low water pressure. If these valves are not adjusted to the proper settings, they can reduce water to little more than a trickle.

Sometimes the adjustment screws on pressure-reducing valves are turned down too tightly. If you are dealing with a low-pressure situation where a pressure-reducing valve is involved, check the level of the screw setting. Turning the adjustment screw counterclockwise will increase the water pressure on the system.

Pressure-reducing valves are not often the cause of a pressure problem, but they can be. It may be that the adjustment screw is set improperly, or it could be that the entire valve is bad. If you cut the water off at the street connection, you can check the building pressure by removing the pressure-reducing valve and installing a pressure gauge on the piping. This will eliminate, or confirm, any doubts you may be having about the pressure-reducing valve.

Galvanized pipe

If you have been in the plumbing trade long enough to have worked with much galvanized pipe, you know how badly it can rust, corrode, and build up obstructions. Any time you are working with a water distribution system that is made up of galvanized pipe, you shouldn't be surprised to find low water pressure.

I have cut out sections of galvanized pipe where the open pipe diameter wasn't large enough to allow the insertion of a common drinking straw. When the open diameter of the pipe is constricted, water pressure must drop. Any long-time plumber will tell you that galvanized water pipe is a nightmare waiting to happen.

Most plumbing systems plumbed with galvanized pipe are equipped with unions on various sections of the piping. If you loosen these unions, you are likely to find the cause of your low pressure. A quick look at the interior of the pipe will probably be all it will take to convince you.

The only real solution to low pressure from blocked galvanized pipe is the replacement of the water piping. This can be a big and expensive job, but it is the only way to solve the problem positively.

Undersized piping

Undersized piping can create problems with water pressure. If the pipes delivering water to fixtures are too small, the fixtures will not receive an adequate volume of water at a desired pressure. As a matter of fact, I was on just such a job earlier today.

I was called to a meeting hall to determine why their water pressure was not as good as the pressure in neighboring buildings. The building is served by a municipal water supply, so that narrowed the list of possibilities.

When I got to the job, I went into the kitchen to test the pressure. Before I ever made it to the kitchen sink I saw the problem.

The water heater for the building is located in the kitchen, in plain view. When I looked over at the heater, I almost couldn't believe what I was seeing. The piping at the heater from both the inlet and outlet connections was up to code, but once the pipes rose a foot or so above the water tank, all bets were off. The $\frac{3}{4}$-inch copper tubing was connected, with flare fittings, to $\frac{1}{4}$-inch copper tubing.

The $\frac{1}{4}$-inch tubing provided the incoming water to the water heater and carried the hot water out to the building's plumbing fixtures. I removed a few ceiling tiles and traced the path of the tubing.

This building is equipped with two bathrooms and a kitchen, and every fixture is supplied with water from the $\frac{1}{4}$-inch tubing. Now I'm not talking about $\frac{1}{4}$-inch branch feeds; I'm saying that all of the water distribution pipe is run with $\frac{1}{4}$-inch tubing.

It is obvious that fixtures designed to be fed by $\frac{3}{4}$- and $\frac{1}{2}$-inch tubing will not perform as well when their supply is reduced to a quarter of an inch.

In a case like this, the only option is to replace the illegal water piping with tubing that will meet code requirements.

Many times the downsizing of piping will not be as drastic as the job I've just described, but inferior piping is common in many rural areas. If you are responding to a complaint of low water pressure, check the pipe sizing.

Hard water

Hard water can be a cause of reduced water pressure. The scale that hard water allows to build up on the inside of water pipes, fixtures, and tanks can reduce water pressure by a noticeable amount. When you have a job that is giving you trouble with low pressure and no sign of a cause, inspect the interior of some piping to see if it is being blocked by a scale buildup.

Clogged filters

Clogged in-line filters are notorious for their ability to restrict water flow. People have these little filters put in to trap sediment, and they do. They trap the sediment so well that they eventually clog up and block the normal flow of water, thereby reducing water pressure.

When you respond to a low-pressure call, ask the property owner if any filters are installed on the water distribution system. If one is, check the filter to see if it needs to be replaced.

Stopped-up aerators

One of the most common, and simplest to fix, causes of low water pressure is stopped-up aerators. The little screens in the aerators stop up frequently when a house is served by a private water supply. Iron particles and mineral deposits are usually the cause for stoppages.

Aerators can be removed and will sometimes clean up, but many times they must be replaced. As long as you have an assortment of aerators on your service truck, this is one problem that is easy to find and to fix.

Too Much Water Pressure

Too much water pressure in a water distribution system can be dangerous and destructive. When fixtures are under too much pressure, they do not perform well over extended periods of time. The O-rings, stems, and other components of the plumbing system are not usually meant to work with pressures exceeding 80 pounds per square inch.

Extreme water pressure can be dangerous for the users of the plumbing system. For example, I once worked for a homeowner who had cut herself badly at the kitchen sink as a result of high water pressure.

The woman was holding a drinking glass under the kitchen faucet when she turned the water on to fill the glass. The water rushed out with such force that it knocked the glass out of her hand. The glass shattered on impact with the sink and pieces of the broken glass sliced into the woman's hand and arm.

High water pressure is easy to control. Water pressure can be reduced with a pressure-reducing valve and, in the case of well systems, with adjustments to the pressure switch.

Troubleshooting high water pressure is simple; all you have to do is look at the pressure gauge on the well system or install a pressure gauge on a faucet, usually a hose bibb. If the pressure is higher than 60 pounds per square inch, it should normally be lowered. Most residential properties operate well with a water pressure of between 40 and 50 pounds per square inch.

If you have a well system, all you have to do is to go into the pressure switch and adjust the cutout setting. In the case of city water service, you will have to install a pressure-reducing valve. If a pressure-reducing valve is already in place, you can lower the system pressure by turning the adjustment screw clockwise.

Water Leaks

Water leaks are probably the most common type of complaint plumbers receive about water distribution systems. Most of the time leaks are easy to find, and they are not usually too difficult for experience plumbers to repair. There are times, however, when the leaks are not so easy to find or to fix.

As the list of acceptable plumbing materials for water distribution systems has grown, so has the diversity of the types of piping and tubing used in water systems. You never know when you will be dealing with copper, polybutylene (PB), polyethylene (PE), chlorinated polyvinyl chloride (CPVC), galvanized steel, or maybe even brass. Just having enough variety of fittings and supplies on your service truck to deal with all of these types of materials can be a job.

Since there are so many possibilities for the makeup of a water distribution system, let's take the time to look at each type of material on its own. This will allow us to see exactly what problems are unique to the different types of pipes.

Copper leaks

Copper leaks are the most common type of leak found in the piping of water distribution systems. Copper tubing and pipe has been used for many years, and it is still quite popular for new installations.

While any experienced plumber knows how to solder joints on copper pipe, there are times when the job doesn't go by the book. A little water trapped in a copper pipe can make soldering a joint a hair-pulling experience if you don't know how to handle the situation. There are also those times when the fittings that must be soldered are located in places where the flame from a torch poses some potentially serious problems.

This book is not designed to tell you exactly how to fix problems on a step-by-step basis. Since the expected readers are professionals, it is assumed that they will know the basics of making standard repairs. However, since some soldering situations are problems in themselves, we will cover some tricks of the trade to help you out in difficult times. After all, this is a problem-solving book.

Finding leaks in copper

Finding leaks in copper pipe and tubing is not usually difficult. By the time plumbers get called in to fix a leak, someone normally knows where the leak is. There are times, though, when the location of the leak is not known.

In other chapters throughout the book we have talked about ways to find hidden leaks. You've seen how to find leaks in the risers to shower heads and how to track down water that is running across a ceiling. Those tactics can be applied to leaks in the water distribution system.

Big leaks are easy to find. You can either see or hear them. It is the tiny leak that does its damage over an extended period of time that is difficult to put your finger on. With these leaks it is sometimes necessary to start at the evidence of the leak and work your way back to the origin of the water. This can mean cutting out walls and ceilings, but there are times when there just isn't any other way to pinpoint the problem.

Copper leaks often spray water in many directions. This can also make finding the leak difficult. If a solder joint weakens to the point that water can spray out of it, the water may travel quite a distance before it splashes down. These leaks are not hard to find when they are exposed, but if they are concealed, you can find yourself several feet away from the leak when you cut into a wall or ceiling that is showing water damage.

The good thing about spraying leaks is that you can usually hear them once you have made an access hole. This is a big advantage over trying to find a mysterious drainage leak. The spraying water will lead you to the leak in a hurry.

Copper pipe and the joints made on it can leak for a number of reasons. Pinholes in the pipe can occur from acidic water. The pipe can swell and split or blow out of a fitting if it has frozen. Stress can break joints loose, and sometimes joints that were never soldered properly will blow completely out of a fitting. Bad solder joints can also begin to drip slowly.

Fixing copper leaks

Fixing copper water lines is usually not a complex procedure, but what will you do if there is water in the lines? Water in the piping can be of two types; it can be standing water that is trapped, and it can be moving water that is leaking past a closed valve. In either case, the water makes it hard to get a good solder joint.

Inexperienced plumbers often don't know how to overcome the problem of water in the piping that they need to solder. Some inexperienced plumbers will try to make a solder joint and be fooled by the actions of the solder. This type of situation will result in a new leak, but the leak may not show up immediately; let me explain.

When water is present in the area being soldered, several things can happen. If there is enough water in the pipe, the joint will not get hot enough to melt solder. This is frustrating, but at least the plumber is aware that the solder joint can't be made without some type of action being taken against the standing water.

In some cases the pipe and fitting will get hot enough to allow solder to melt, but the fitting will not obtain a temperature suitable for a solid joint. When this happens, solder will melt and roll out around the fitting, but it is not being sucked into the fitting as it should be. To inexperienced eyes, this type of joint can look okay, but it's not.

When the water is turned back on, the defective joint may leak immediately, if you're lucky. If the joint leaks right away, the plumber knows the job is not done. However, sometimes these fouled joints will not leak immediately, and this means trouble.

When the weak joint doesn't leak during the initial inspection, it may be left and possibly concealed by a wall or ceiling repair. In time, and it usually won't be long, the bad joint will begin to leak. It may drip, or it may blow out. Either way, the plumber's insurance company won't be happy.

When the bad joint fails and is inspected by the next plumber, it will be obvious the joint was not made properly. There will not be evidence of solder deep in the fitting because it was never there. This will usually be considered neglectful on the professional plumber's part.

A third way that water in the piping will drive an inexperienced plumber crazy is with the results of steam building up in the pipe. As the area around the joint is heated, the water will turn to steam. The steam will vent itself, usually through some portion of the fitting being soldered. This steam may escape without being seen. When this happens, solder runs around the fitting as it should, except that a void is created where the steam is blowing out. If the plumber can't see, hear, or sense the steam, the joint will look good. Once the water is turned on, the joint will no longer look so good; it will leak and the process will have to be repeated.

Inexperienced plumbers will think they just had a leak because of poor soldering skills. They will go back through the same process and wind up with the same results. Until they figure out what is happening, and what to do about it, they will just spin their wheels trying to solder around the steam. There are ways to beat all of these problems, and I'm about to show them to you.

Trapped water.   Trapped water is easier to overcome than moving water. Trapped water can sometimes be removed by bending the cut ends of horizontal pipes down. Opening fixtures above and below the work area will often remove trapped water from pipes. When the problem pipe is installed vertically, a regular drinking straw can be inserted into the pipe and the water blown out. If you have the equipment and time, you can use an air compressor to blow trapped water out of pipes. When none of these options work, you have to get a bit more creative.

The easiest way to beat trapped water in pipes that just won't drain is the use of bleed fittings. These are cast fittings that are made with removable drain caps on them. Heating mechanics call them vent fittings, I call them bleed fittings, and many people just call them drain fittings.

Drain fittings are available as couplings and ells (Figs. 17.2 and 17.3). The vent on a coupling is on the side of the fitting, at about the center point. Drain ells have their weep holes on the back of the ell, where the ell makes its turn.

If you install a drain fitting in close proximity to the trapped water, you can steam the water out of the pipe as you solder the fitting. Remove the cover cap and the little black seal that covers the drain opening. Make sure the drain opening is not pointing toward you.

As you heat the pipe and fitting, the trapped water will turn to steam and vent through the weep hole in the fitting. The hot water can spit out of the hole and the steam can come out fast and hot. This is why you don't want the hole pointed in your direction.

With the water and steam coming out of the vent hole, you can sol-

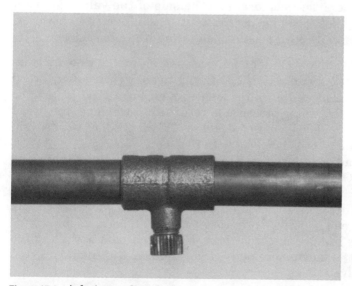

Figure 17.2 A drain coupling.

Figure 17.3 A drain ell.

der the joint with minimal problems. Unless you have a huge amount of water in the pipe, the soldering process shouldn't take long, and it should go about the same as it would if the water wasn't in the pipe.

Once the joints have cooled, you can replace the black seal and the cap to finish your watertight joint. If you get caught without bleed fittings on your truck, a stop-and-waste valve can be substituted for the drain fitting. The little weep drain on the side of the valve will work in the same way described for the drain fittings. Make sure, however, that the valve is in its open position before you begin to solder.

Bread. Bread is an old standby for seasoned plumbers who are troubled by water in the pipes they are trying to solder. Inserting bread into the pipe will block the water long enough for a good joint to be soldered. When the water pressure is returned to the pipe, the bread will break down and come out of a faucet. This procedure works well, but there are a few traps involved with it.

I've used bread countless times to control water that was inhibiting my soldering. On occasion, I've regretted it. Sometimes I've created more problems for myself by using bread. I would recommend that you use drain fittings rather than bread, but I'll tell you what to look out for when bread is used, just in case you don't have a choice.

First, always remove the crust from loaf bread before stuffing the pipe. The crust is much more dense than the heart of the bread, and it doesn't dissolve as well.

When bread has been used in a pipe, try to remove it through the spout of a bathtub. If you must get it out through a faucet, remove the aerator before you cut the water on. The broken-down bread will clog the screen of the aerator in the blink of an eye.

Don't flush a toilet to remove bread from the line. The bread particles may become lodged in the fill valve and cause more trouble.

Avoid packing the bread in the pipes too tightly. I once used a pencil to push and pack bread into a pipe that had moving water in it. The bread blocked the water and allowed me to make my solder joint, but it didn't come out once the water was turned on. Oh, it came through after awhile, and blocked up the supply to a toilet.

Moving water. Moving water makes the job of soldering joints an adventure. Old valves don't always hold water back completely. There will be many times when a trickle, or more, of water will creep past the valve and make soldering a bad experience.

When you have moving water in a pipe, you will have to use some special way to get the soldering job done. Bleed fittings will let you do your work if the water is not too abundant.

There is a special tool that I've seen advertised, but I can't remember its name, that is designed to make soldering wet pipes possible. As I recall, the tool has special fittings that are inserted into the pipe and expanded. The plug holds back the water while a valve is installed on the pipe. When the valve is soldered onto the end of the pipe, the tool is removed and the valve is closed, allowing you to work from the valve without the threat of water.

The tool is nifty but not needed. You can do the same thing with the old standby, bread. Stuff the pipe with bread, and pack it if you have to. Once the water is stopped, solder a gate valve onto the end of the pipe. Don't use a stop-and-waste valve because you may need access to the bread in order to get it out.

Once the valve is soldered properly and cooled, close it and cut the water on. Open the valve to let the bread out of the system, and remember that water will be coming out right behind the bread. If the bread is packed too tightly for the water to come through, poke holes in it with a piece of wire, like a coat hanger. The bread will break up and come out of the pipe. Once the pipe is clear, close the valve and go on about your business.

Use a heat shield. When you are soldering in close quarters, use a heat shield and have a fire extinguisher close by. If your work is dangerously close to combustible materials, wet them down with water before putting your torch into the area.

Heat shields that attach to the tip of a torch are available, but these units are small and are not always enough to consider the conditions safe. I carry a piece of duct work on my truck that can be used to block flames from combustibles, and I've used aluminum foil in years past. Both work well, but you should never depend too heavily on any heat shield to provide positive protection; keep a fire extinguisher at hand.

I knew a plumber once (he didn't work for me) who almost set an entire apartment building on fire. He was repairing a copper pipe that had split due to freezing. During his work, the insulation in the wall caught on fire. The plumber had limited experience, and he tried to put the fire out at the point where it had started. The wall, however, was acting as a chimney, and the fire was spreading upward, quickly.

The plumber's boss was in the next room making some repairs and came running in. As soon as he saw what was happening, he used his hammer to open the wall where it met the ceiling, several feet above the work area. I understand it was touch and go, but the supervisor was able to contain the fire in the one apartment and get it put out.

The apartment building was old and not built to present-day fire codes. If the experienced supervisor had not been on the job, the whole building probably would have gone up in flames. Keep this story in mind, and keep that fire extinguisher close by. And remember to head fire off if it is spreading, not to fight it just where you can see it.

PB leaks

PB leaks are not common, except at connections that were not made properly. Because of its flexibility and durability, PB pipe rarely gives plumbers problems with leaks. It doesn't even normally burst under freezing conditions. There are, however, several times when the pipe is not put together properly, and this can result in leaks.

Bad crimps. Bad crimps account for some leaks with PB pipe. If a crimping tool gets out of adjustment, and they do after extended use, the crimp ring may not be seated properly. If this is the case, the bad connection must be removed and a new one made.

Fittings not inserted. It is not often that an insert fitting is not inserted far enough, but I've seen it happen. This type of problem is obvious and normally not difficult to repair. The replacement of the bad connection is all that is required.

Stainless steel clamps. Some plumbers are not familiar with PB pipe, and they sometimes use stainless steel clamps to hold their connections together. This rarely works for long, if at all. If you have a leak at a PB connection where a stainless steel clamp has been used, remove the clamp and crimp in a proper connection.

Compression ferrules.   Compression ferrules may account for most of the after-installation leaks in PB piping systems. Compression fittings can be used on PB pipe, but brass ferrules should be avoided (Fig. 17.4). Nylon ferrules are the proper type to use with compression fittings and PB pipe.

If you have a PB pipe or supply tube leaking around the point of connection where a compression fitting has been used, inspect the ferrules. If brass ferrules have been used, replace them with nylon ferrules. The brass ferrules can cut into the PB pipe if the compression nut is tightened too much.

Figure 17.4   A polybutylene closet supply.

I'm not saying that brass ferrules can't be used on PB pipe. In fact, I've used them on PB supply tubes many times without any problems. But, if the plumber puts too many turns on the compression nut, the PB will be cut. Whenever possible, use nylon ferrules.

PE leaks

PE leaks are not much of a consideration when talking about water distribution systems. PE pipe is used frequently for water services, but its use is very limited in water distribution systems.

Cracked fittings and loose clamps are the most frequent causes of leaks in PE piping. The fittings can be replaced, and the clamps can be tightened or replaced.

On the occasions when PE pipe develops pinhole leaks, there are a few easy ways to fix the problem. The best way to deal with the situation is to cut out the bad section of the piping and replace it. When this isn't convenient or possible, you can use repair clamps.

Repair clamps can be used to patch small holes in all types of piping. You can buy the clamps, or in some cases you make your own. In the case of PE pipe, rubber tape and a stainless steel pipe clamp will make a good repair on small holes. All you have to do is wrap rubber tape around the hole and clamp it into place.

CPVC leaks

CPVC leaks are common, and in my opinion, are a real pain to deal with. I've worked with CPVC off and on for going on 20 years, and I've never liked it. The one house that I plumbed with CPVC was enough to turn me off from using it in future jobs. Since that first house, early in my career, my only association with CPVC has been in repairing it.

If you have worked with CPVC much, you know it is brittle, takes a long time for its joints to set up, and is famous for its after-the-fact leaks.

When you are troubleshooting a water distribution system made from CPVC piping, there is a lot to look for. You have to keep your eyes open for little cracks in the pipe. Because of its makeup, CPVC will crack without a lot of provocation. These hairline cracks can be hard to find.

In addition to cracks in the piping, you have to look for cracks in the threaded fittings. Fittings that have been cross-threaded are also common in CPVC systems. And to top it all off, CPVC pipe that is not supported properly can vibrate itself to the point of weakening joints that will leak.

When you find a leak in CPVC pipe or fittings, don't attempt to reglue the fitting. Cut it out and make a good connection with new materials. Attempting to patch old CPVC will more often than not result in frustration and continued problems.

You don't have to solder CPVC, but water in the line will still mess up your new joints. The water will seep into the cement and create a void that will leak.

I know some plumbers just cut the pipe, slap a little glue on it, and stick it together with its fitting, but I don't think you should operate this way. CPVC is so finicky that I believe you should make your connections by the book.

Cut the ends of the pipe squarely and rough them up with some sandpaper. Using a cleaner and a primer prior to applying the cement is also a good idea. When you glue and connect the pipe with the fitting, turn it if you can to make sure the cement gets good coverage.

Don't turn the pipe loose right away. If you release your grip too soon, the pipe is likely to push out of the fitting to some extent, weakening the joint. Hold the connection in place as long as your patience will allow. CPVC takes a long time to set up, so don't move it or cut the water on too soon. New joints should set for at least an hour before being subject to water pressure.

Galvanized steel and brass

Galvanized steel and brass pipes aren't used much for water distribution these days, but they do still exist in some buildings. These types of pipe present some special problems.

Leaking threads.   Leaking threads are a common problem with old piping. The threads are the weakest point in the piping, and they tend to be the first part of the system to go bad. Acidic and corrosive water can work on these threads to make them leak prematurely.

When you have a leak at the threads going into a fitting, a repair clamp is not going to help you. Under these conditions, the leaking section of pipe must be removed and replaced. This can become quite a job.

What starts out as a single leak at one fitting can quickly become a plumber's nightmare. As you cut and turn on the bad section of pipe, you are likely to weaken or break the threads at some other connection. This chain reaction can go on and on until you practically have to replace an entire section of the water distribution system.

Many young plumbers see leaks at threads and believe they can correct the problem by tightening the pipe. In old piping, this is only a pipe dream, no pun intended. Twisting the old pipe tighter is not likely to solve problem, but it may worsen it.

If you have leaks at the threads, you might as well come to terms with the fact that the section is going to have to be replaced. Unlike drain pipe, you can't use rubber couplings to put water piping back together. You have to remove sections of the pipe until you can get to good threads. Then you can convert to some type of modern piping to replace the bad section. Be prepared to have to replace more than you plan to.

Pinholes.   Pinholes in these hard pipes can be repaired with repair clamps. Rubber tape and a pipe clamp will work, but a real repair clamp is best.

Brass imitates copper.   Brass pipe imitates copper to the point that some plumbers will think they are working with copper. In fact, I hate to admit it, but I've even mistaken brass for copper. It is easy to do in poor lighting.

Brass water pipe can be cut with copper tubing cutters. The cut takes a little longer, but unless you suspect the pipe is brass, it can fool you. However, once you try to get a copper fitting on the end of the pipe, you'll know something is wrong; the fitting won't fit.

If you think you're working with copper and can't figure out why your fittings won't work, look for an existing joint and see if it is a threaded connection. If it is, you're probably dealing with brass pipe.

If the fittings are soldered and your standard fittings won't fit the pipe, somebody plumbed the job with refrigeration tubing. When this is the case, you're going to need refrigeration fittings to get the job done. This one tip can save you a lot of frustration.

Compression leaks

Compression leaks are frequent problems with water distribution systems. The compression fittings are usually easy to fix. All that is normally required is the tightening of the compression nut. These leaks happen when the nuts were not tight to begin with or when the pipe has vibrated enough to loosen the fitting. It is also possible that the connection was hit with something and loosened.

I responded to a call last week that would fool a lot of plumbers, so I think it is worth telling you about. The call came in from the owner of a motel. His maintenance mechanic had investigated a leak under a lavatory and determined that a plumber should be called in. The owner explained to me that the leak was at the threads of a straight stop under the lavatory. He said the leak was below the cutoff and the water supply to the whole motel would have to be cut off.

I went to the job personally and inspected the leak. The piping came up through the floor. A long, chrome nipple ran from the floor to the straight stop. A steady stream of water was running down the nipple, and at first glance, it did appear to be coming from the threaded connection.

Both the owner of the motel, who is handy, and the maintenance person had inspected the leak. They were convinced the problem was with the threaded connection. If I had wanted to take advantage of the customer, I could have easily played along with their opinions and made a little job into a much larger one.

I could have inconvenienced everyone in the motel by cutting off the main water supply and worked for some time in replacing the nipple and the cutoff valve. Instead, I did the right thing.

What did I do? I didn't rely on the information I had been given. I looked at the problem closely, as any good troubleshooter would, and I saw what I believed to be the real problem. Do you know what it was?

When I looked at the stop valve, I saw a bubble of water sitting on top of the compression nut that held the supply tube in place. I cut the supply valve off and wiped all the water off the nipple. The water didn't continue to run down the nipple. Obviously, if the leak had been at the threads on the bottom of the valve, the leak would have continued.

I had been reasonably sure that the leak was at the supply-tube connection, but my little test proved me to be right. I tightened the compression nut on the supply tube and cut the valve back on. The leak was gone.

Instead of spending an hour or so replacing the nipple, only to find that it wasn't the problem to begin with, I completed the job in less than 10 minutes.

The maintenance person was a bit embarrassed by my ease in fixing the problem, and the owner was delighted. They had both been nearby during the repair and knew that I had not cut the main water supply off or spent much time on the job.

They asked how I had fixed the problem so quickly, and I told them. While I was on the job, the owner had me rebuild a tub valve and replace a relief valve in a water heater. My honesty, professionalism, and troubleshooting skills were appreciated so much that I now have the account not only for that motel but for another one as well. The owner has even asked me to plumb 10 new units he will be building in a month or so. It is surprising what a little job can lead to when you do it right.

I can easily understand how the two people thought the leak was at the threads. Even a plumber who was in a hurry might have assumed the same thing. By troubleshooting the job properly, the root cause was found and the job was simple to do. This is well worth remembering in your work.

Frozen Pipes

Frozen pipes can be a plumber's bread-and-butter money in the winter. They can also be troublesome to work with, hard to find, and difficult to fix and are potentially dangerous if they are not worked with in the proper manner.

Steel pipe, and sometimes copper pipe, can be thawed with the use of a welding machine. If the leads are attached at opposite ends of the pipe, with the frozen section in the middle, the welding rig can produce enough juice to thaw them. I've seen plumbers do this many times over the years, but the process can be dangerous; fires can be started.

Special thawing machines are also available for dealing with frozen pipes. They work on pretty much the same principle as the welding machine. I prefer to avoid using electricity to thaw frozen pipes.

When I was a plumber in Virginia, I used to get numerous calls for thawing and repairing frozen pipes. Normally, the job entailed only a single pipe, often that of an outside hose bibb. A heat gun, a hair dryer, or a torch made quick work of thawing these individual sections of pipe.

In Maine, the freeze-ups are often considerably larger than the ones I dealt with in Virginia. Here it is not uncommon for whole systems to freeze. To thaw these pipes, I usually use a portable heater. A large space heater, like those used on construction sites, will bring a building up to a thawing temperature quickly and safely. Once the building is warm and the pipes are thawed, necessary repairs can be made.

If you haven't worked with many frozen pipes, you may not be aware of the way the freezing action can swell copper pipe. Even though the split resulting from the freeze-up is in one place, the pipe might be swollen for several feet on either side of the split. This makes it impossible to get a fitting on the end of the swollen pipe. To handle this dilemma, you will have to keep moving back on the pipe until you find a piece that has not swollen from the cold.

Knowing What to Look For

Knowing what to look for is a key aspect of any type of troubleshooting. If you know what to look for, you have a much better chance of finding it. Problems with water distribution systems can involve many possibilities, but they are not hard to work through when you use good troubleshooting skills. This chapter has shown you the ropes on water pipes, so let's move onto the next chapter and see what's in store for us there.

18

Troubleshooting Drainage and Vent Systems

Troubleshooting drainage and vent systems can involve a lot more than just looking for leaks and stoppages in the drainage system. While leaks and stoppages are the two most common complaints with drainage, waste, and vent (DWV) systems, they are far from being the only problems people experience with their DWV systems.

Let me give you a quick quiz to see how much you already know about troubleshooting DWV systems. I'll pose the questions to you now and answer them as we move through the chapter. Get a note pad and jot down your answers to the questions. We'll see how you did you did on the quiz a little later.

1. This question deals with odors in a home. This house has three bathrooms, two with tub/shower combinations and one with a stall shower. The house has two stories with two baths upstairs and one, the one with the shower, downstairs. The homeowner has noticed that when she is working in her kitchen there is a bad odor in the area. There are several possible causes for this problem, but what would you look for first?

2. This question has to do with a noisy drain pipe in the wall between the kitchen and the living room of a home. The homeowner despises having guests listen to water rush down this drain every time the upstairs toilet is flushed. He is willing to go to any reasonable expense to correct the problem. What would you suggest?

3. The third question in our quiz has to do with the drainage from a commercial dishwasher. The business owner calls you and explains that the plastic pipe that carries the drainage from his dishwasher is

leaking again. He goes on to explain that this is the third time he has had the same problem, and he wants you to fix the piping so that it won't happen again. What will you do?

4. This one has you looking for the reason why the fixtures in the bathroom of a new home are not draining properly. The bathtub and the lavatory in this new house are both draining slowly. You have snaked the drains, but they still won't create the drainage whirlpool that they should. What might the problem be in this case?

5. This question finds you wondering why a kitchen sink that had been working satisfactorily is suddenly backing up. All the homeowner can tell you is that a garbage disposer was recently installed on the sink, and that seems to be when the problem started. You should be able to make a good guess about the cause of the problem without even looking under the sink. What do you think the problem is?

6. A homeowner has called you and asked if it is possible for their vent pipes to freeze in the winter time. How will you answer the customer?

7. This question has to do with a foul odor in the basement of a home. The basement contains most of the home's DWV system. There is also a floor drain and a sump for a sump pump in the basement. What is the most likely cause of the foul odor?

8. This question has to do with a kitchen drain that stops up frequently. The homeowner has explained to you that she has to have the drain snaked out about every month. It seems the plumbers always hit the clog within 15 feet of the sink. The fixture drains okay for a while, but then it stops up again. What do you suspect the answer to this problem is?

9. This question has to do with a homeowner who went into his attic to get down some items he had in storage. While the man was in the attic, he noticed a strange smell. At first, he thought maybe a small animal had died in the attic, but he couldn't find any evidence of it, so he called you. What are you going to look for first?

10. The final question in our quiz revolves around a house with old plumbing in it. The house has a half-basement under it. The rest of the foundation, where the building drain passes into it, is a tight crawl space. The building drain of the house is stopped up. The old cleanout in the basement doesn't want to come out, so you elect to snake the drain from the first-floor toilet. You send your snake down the drain, and it hits something solid. The snake cable kinks and won't go any further. The cable is acting like it is hitting a ball of tree roots or a broken pipe.

When you can't get the snake to go in any further, you retrieve it and note how many feet of cable you had in the drain. After going into the basement and guesstimating the distances of the piping, you believe the snake was hung up in the pipe section that is in the crawl space. Since the problem is in the crawlspace, you rule out the possibility of tree roots. A quick inspection with a flashlight seems to rule out a collapsed pipe. What else might the problem be?

I'm about to give you the answers to the above questions. Check your notes to see how you did. If you did well, congratulations. Should your score be lower than you would like, at least you learned your lessons here, instead of in the field.

1. The most appropriate answer for question 1 is a dry trap in the downstairs shower. This question was based on a real-life problem that my mother-in-law was experiencing. She asked me one day if I had any idea where the odor in her kitchen was coming from. I knew she had a shower just down the hall from the kitchen, and I didn't think she used it often. After asking if the shower got much use, she confirmed my suspicion. Acting on my instructions, she ran the shower for a while to replenish the seal in the trap. The problem has never come back.

2. Many people don't like having noisy drains in the vicinity of their living rooms. Unfortunately, there is nothing simple that can be done about this problem once the pipes are installed and the walls are finished off.

Since this homeowner is willing to go to some expense, there are two viable options. The plastic drain could be replaced with cast-iron pipe. This would help to deaden the sound. Insulation could then be wrapped around the cast-iron pipe to muffle it even more. It would also be conceivable to just wrap insulation around the existing drain. Of course, the wall with the pipe in it would have to be opened and repaired to allow for this type of work.

3. The best way to solve the problem in question 3 is to replace the plastic pipe with either DWV copper or cast-iron pipe. The intense heat from the commercial dishwasher is more than the plastic pipe and associated joints can handle.

4. The answer to question 4 is one that I doubt many got right. If you missed this one, don't feel bad. I have seen the problem occur, but the reason for it is one that should not exist. What was the cause of the problem? The plumber who installed and tested the new piping forgot to remove the test caps from the vents on the roof. This may sound weird, but I've run into the problem twice.

5. I hope you got question 5 right. Why would the drain suddenly be backing up? My guess would always be that the kitchen drain was plumbed with old galvanized pipe. The pipe was probably closing up with rust, grease, and assorted gunk. I expect there was a small hole in the center of the closing pipe that would allow water from the sink to drain, but with the disposer on the pipe, the small opening left in

the drain pipe couldn't handle it. This type of problem happens so often that I won't install a garbage disposer on a sink that drains into galvanized pipe unless the property owner releases me from liability for stoppages.

6. Is it possible for vents to freeze as was asked in question 6? Yes, it is possible. In extremely cold conditions, vents can condensate and the moisture can freeze on the inside of the pipe. If the temperatures stay low for a long time, the ice can eventually reach all the way across the pipe and seal the vent.

7. How did you answer question 7? If you guessed that the trap for the floor drain had dried up, you're right. Water in traps will evaporate over time. If the drains are not used periodically, the water seal will dissipate and odors can escape. In some cases, trap primers must be installed to make sure the condition does not occur.

8. After reading question 8, did you guess the problem to be galvanized piping that is closing up? If you did, you're right on the money. As galvanized drains close up and are snaked out, generally only a small hole is punched through the obstruction. Since few plumbers run cutting heads down sink drains, their spring heads do nothing more than open small passages in the drain. This works for a while, but then new grease and gunk joins the existing obstruction to block the pipe again.

9. What do you think the problem is in question 9? More than likely the problem is a vent pipe that was never extended through the roof. It could be a pipe that was broken after installation, but it's probably just one that didn't get hooked up. If you are faced with this type of situation and don't see any pipes that are not installed properly, look under the insulation. Many times the insulation will be covering the pipe that is responsible for the problem.

10. If you're a young plumber, you may have had a problem with question 10. Older plumbers probably knew right off what to expect. The problem would probably be an old house trap.

In the old days, it was common to install a house trap in the building drain or sewer in close proximity to where the drain left the foundation of the property. Sometimes the traps are installed inside the foundation, and sometimes they are installed underground, outside of the foundation. When they were used, house traps were usually the only traps installed on the drainage system. Large snakes frequently can't get through these traps. If you have a mystifying problem with your snake in an unusual section of drainage pipe, you can make a pretty safe bet that you will find a house trap installed.

Now that you've had a chance to test your knowledge of DWV systems, let's move on into the chapter and see what else you can learn.

Clogged Vents

Clogged vents can cause fixtures to drain slowly. You've seen one example where the test caps were never removed from a home's vents, but that is not the only way that vents get plugged up.

Squirrels and birds can disable a plumbing vent very quickly, especially vents with small diameters. Birds sometimes start building nests on top of vent pipes, and squirrels have been known to use the pipes as a cache for nuts.

Aside from wildlife, Mother Nature can put a strain on plumbing vents. If a vent is positioned under trees, all sorts of things can get into the pipe. Leaves can fall into the vent, pine cones could plug the pipe, and a combination of nuts, leaves, and twigs can restrict the opening of the vent pipe.

If you have a drain that is draining slowly after a thorough snaking, you should check to see first if the fixture is vented. If it is, check the vent for stoppages. If necessary, you can snake the vent from the roof to break up stoppages.

When a fixture is not vented, you might consider installing a mechanical vent on the fixture to allow it to drain better. Some code officers frown on this practice, so make sure you are within the limits of code compliance before you put a mechanical vent on the system.

Vents Too Close to Windows

Vents that are too close to windows that open, doors, or roof soffits can be responsible for odors in a building. The vents should be at least 10 feet away from such openings. If the vents are closer than that, they should extend at least 2 feet above the opening. Otherwise, the fumes from the vent pipe can be sucked into the building through the opening.

Frozen Vents

You learned in the quiz questions that vents can freeze. Many plumbers don't realize this. When a plumbing system suddenly fails to drain as well as it normally does in cold weather, look to the vents to see if they are frozen. If they are, an acid-based drain cleaner will normally eat through the ice and clear up the problem, at least temporarily.

Vents in General

Vents in general don't present many problems for plumbers. Unless they are nonexistent or stopped up, vents don't normally require any attention.

Copper Drainage Systems

Copper drainage systems are usually not much trouble. The copper provides many years of good service, and it is not normally a contributor to stoppages. In fact, in 20 years, I can't recall a time when a copper DWV system caused any problems. Oh sure, they get stopped up, but all drains can. In general, copper is a good above-ground drainage material that is very dependable.

Galvanized Pipe

Galvanized pipe has to be one of the worst materials ever used in plumbing. This pipe is famous for its ability to rust and to catch every imaginable thing that goes down it. Grease and hair are especially common stoppages found in galvanized drains.

As galvanized pipes age, they begin to develop buildups that slowly restrict their openings. Eventually, the buildup blocks the pipe completely. Snaking the drain will punch holes in the obstructions, but the stoppage will recur in a matter of weeks or months.

Old galvanized drains are also known to rust out at their threads, causing leaks at their joints. The best solution to galvanized drain problems is the replacement of the old piping with more modern plumbing materials. Rubber couplings make it easy to adapt new drainage materials to the old piping.

Cast-Iron Pipe

Cast-iron pipe has been used for DWV systems for longer than I've been a plumber and then some. Typically, cast iron gives good, dependable, long-term service. Whether the system is plumbed with service weight pipe or no-hub pipe, the cast iron lasts a long time.

There are some drawbacks to cast-iron drains, however. The interior of cast-iron pipe can be rough, especially as it starts to rust. These rough surfaces catch a lot of debris as it is being drained down the pipes. This leads to stoppages, but they are not as bad as the ones found in galvanized pipes, and they don't recur as quickly.

One of the biggest problem with cast-iron drainage systems is the removal of cleanout plugs. The old brass plugs that have been in the cleanouts for years can be next to impossible to remove.

I've had occasions when a 24-inch pipe wrench with a 2-foot cheater bar wouldn't turn the cleanout plug. If you've done much work with pipe wrenches and cheater bars, you know the kind of leverage that is being applied under these conditions. Even so, some cleanout plugs just won't budge, and some of them are in locations where you just can't get enough leverage on them to make the turn.

When this happens, your options are limited. Some plumbers drill the brass plugs out. Others use cold chisels and hammers to knock them out. And others just cut the pipe and put it back together with a rubber coupling to make cleaning the drain the next time easier. I usually opt for the latter.

I had a service call a few weeks ago that I would like to tell you about. The call came in from a customer who owns a few rental houses. One of his houses was suffering from a stopped-up drain. He told me that nothing in the house would drain and that the condition had existed for a few days.

I went to the house and talked with the tenant. Sure enough, all of the fixtures on the main floor of the one-level house were out of service. When I went into the basement, water was leaking through the floor at the base of the toilet. The building drain was made up of cast-iron pipe. Some of the fixture branches were piped with PVC pipe.

As I looked around, trying to decide which drain cleaner to use, I noticed a washing-machine hookup. It was located near where the building drain went through the foundation. I thought it strange that the fixtures above me were all backed up, but no water was coming out of the washing-machine receptor.

The sewer had a 4-inch diameter. The combination wye-and-eighth-bend fitting at the foundation wall reduced the building drain to a 3-inch diameter. The horizontal extension of the wye had been reduced to a 1½-inch diameter and served only the washing-machine receptor. It had a cleanout in the end of it and was plumbed with PVC pipe. Tentatively, I removed the cleanout plug. Nothing came out. I didn't really think anything would, but you can never be sure with plumbing.

The fact that nothing flooded out of the cleanout or the laundry receptor told be that the blockage was not in the sewer, but in the 3-inch building drain.

The building drain had a developed length of about 24 feet and very few offsets. In fact, the only offset was one long-sweep ell. This looked as though it would be a very simple job, but boy was I wrong.

A section of the horizontal drain near the bulkhead door was cracked and leaking. I suspect the crack came from a previous stoppage that froze before it was cleared up. Wind probably whipped through the bulkhead door and froze the contents of the pipe, cracking the cast iron. At any rate, the dripping crack told me that the stoppage was downstream of the broken pipe.

Unfortunately, there was no cleanout at the change in direction of the pipe, only a long-sweep ell. The only cleanout upstream of the clog was a small one, an 1½-inch one, where a branch took a turn for the kitchen sink. This cleanout was plumbed in with PVC, so I decided to try to clear the stoppage through it.

I removed the cleanout plug and caught the backed-up liquid in a bucket. Since the pipe was hanging tight to the floor joists, I decided to use a hand-held snake for my first attempt. I put 25 feet of cable down the drain and the spring head brought back a significant amount of hair and toilet tissue. Hoping the drain was cleared, I had the homeowner run the sink upstairs. It didn't take long for the water to back up and come out the cleanout. Another tactic would be needed.

I don't like to run a big sewer machine in a drain that is over my head, and all of these pipes were high. There was a set of stairs going up near the drain for the lavatory and toilet. This gave me an idea.

There was a short section of PVC that connected the toilet and the lavatory to the cast-iron drain. I could have gone upstairs, pulled the toilet, and snaked down from above, but I wanted to avoid pulling the toilet and making a mess in the bathroom.

I cut the $1\frac{1}{2}$-inch lavatory drain and set my big drain cleaner up on the steps. A paint can was placed on the lower step to provide a solid base for the machine. The snake with a spring head went down the drain easily. It also brought back hair and toilet tissue when it was retrieved.

The tenant ran the upstairs sink again. The drain was still plugged up. I tried various heads on the snake cable, but none of them cleared the stoppage. Confused, I tried a sewer bag with water pressure. Still nothing happened. Next, I put a flat-tape snake down the drain. I hit resistance and broke through it, but the upstairs sink still backed up.

It had been nearly 2 hours since I started working with this simple drain stoppage, and it was still plugged after I had taken my best shots at it. I knew the problem had to be in the horizontal section of the building drain, and I couldn't figure out why it wouldn't clear.

Assuming the landlord would want the cracked piece of drainage pipe replaced, I decided to break it out and snake from that point. I broke out the side of the pipe and was promptly met with all sorts of nasty contents. The stoppage was definitely downstream, but now there was less than 15 feet of pipe for it to be hiding in.

Since the pipe was over my head, I used the flat-tape snake on it. I met with strong resistance and couldn't get the snake through the stoppage. By this time, I'm thinking some child's rubber ball or duck got flushed down the drain.

I was hot, sweaty, and beginning to lose my patience. Never in 20 years had I run across such a stubborn stoppage that was this unexplainable.

I went to where the building drain offset downward to the sewer. There was a short section of pipe between the combination-wye-and-eighth-bend and the eighth-bend that brought the pipe into a horizontal position. Not having a set of cast-iron cutters with me, I broke out this section of the piping with a framing hammer. Nothing came out of the pipe. Aha, the stoppage was between the two open sections of piping. There was no way it could escape me now.

I put a sewer bag in the pipe and turned on the water pressure. Nothing happened. I couldn't believe it. Next, I went back after the clog with the tape snake. Knowing that I was hitting the stoppage and not a fitting, I put all of my considerable weight behind the snake. When the clog broke free, it splattered all over the basement floor, near the sump pump.

If the tape snake had not gotten the clog, I was going to remove the entire section of piping and dissect it to see what was going on, even if it were at my own expense. Though much of the clog blew out into the basement, a lot of it was still in the pipe. The snake pushed a huge amount of unidentified crud out of the pipe. It was clearly the worst residential clog I've ever seen. I don't understand how so much stuff could develop inside a pipe before someone called a plumber. It was as if the whole 3-inch diameter was filled with debris.

Anyway, I replaced the cracked pipe and put the rest of the pipe back together. The problem was solved, but it took nearly 3 hours. The electric drain cleaners were apparently just punching holes through the stoppage. I wasn't able to use a large cutting head because I was accessing the pipe through an 1½-inch drain. The end result was a problem that you won't learn about in apprenticeship classes or by reading most books. It was one of those rare occurrences that in my case has only happened once in 20 years, but it happened. Typically, cast-iron drainage piping doesn't produce the kinds of problems you have just read about. It is normally a good, dependable DWV pipe.

Plastic Pipe

Plastic pipe, as you probably know, is the most common pipe used for DWV systems today. It may be ABS or PVC, but both are good and dependable. PVC is more brittle than ABS and is more likely to be cracked or broken, but under normal conditions, neither pipe gives much cause for trouble.

Drum Traps

Drum traps are illegal for most uses under most plumbing codes. With the exception of combination-waste-and-vent systems, drum traps are hardly ever used. They are, however, prevalent in Maine, where few fixtures are individually vented.

If you are trying to snake a tub or shower drain and your snake begins to kink up quickly, you may be wrapped up in a drum trap. Because of their design, drums traps will not allow a snake to pass through them. You probably won't run into many drum traps, unless you work in Maine, but it does help to know that they may exist and that they impede the progress of a snake.

Lead Traps and Closet Bends

Lead traps and closet bends are not found too often these days, but there are some still in use. If you are called in to a building where a ceiling is showing water stains that are probably drainage related, the problem may be with a lead trap or closet bend.

The lead used in these fittings gives out after awhile, and leaks develop. There is no need to attempt to repair old lead traps and closet bends; it is best to replace them.

Mystery Leaks on the Floor

Mystery leaks on the floor in a commercial building can be caused by indirect wastes that are plugged up. Sometimes indirect waste pipes are blocked almost completely. It would be easy to diagnose the problem if the pipe were blocked completely, but when there is enough of an opening for water to slowly slip by, pinpointing the problem can be more difficult. Let me give you a case-history example about an ice machine in a motel.

I was a field supervisor for a plumbing company when this story took place. A motel manager called the company I worked for and requested service for a leak at her ice machine. A plumber was dispatched, and he found a puddle of water under the drainage piping of the ice machine.

The drain was made of $\frac{3}{4}$-inch copper tubing with sweat joints. The drain terminated over an indirect waste with a $\frac{3}{4}$-inch copper ell dumping into an $1\frac{1}{2}$-inch open drain.

The plumber searched for the leak and couldn't find it. He told the manager to mop up the floor and keep an eye on it. If water reappeared, he would come back. The manager had the floor mopped, and the next day the puddle was back.

The manager called the company and requested a different plumber. Since I was the field supervisor and the public relations person for the company, I was dispatched to the call.

When I arrived there was a shallow, but wide puddle of water on the tiled floor. I inspected the copper drainage tubing and could find no evidence of a leak. The pipe was dry and the fittings all looked to be in good shape and well soldered.

There was no way to tell that the indirect waste was stopped up by looking at it, since the trap holds water at all times. I had never experienced anything like this, but it made sense to me that the only reason for the puddle had to have something to do with the indirect drain.

I wasn't sure if some other fixture was causing the drain to back up and spill its contents or if the puddle was being made by the discharge of the ice machine. There was, however, an easy way to find out.

The motel manager gave me a pitcher full of water that I poured into the indirect waste, and guess what? Yep, the pipe overflowed, creating a puddle. I snaked the drain and the problem never recurred, to the best of my knowledge.

In this case, the occasional dripping of the ice machine was not enough to make the pipe overflow, but when the machine went into its heavy discharge cycle, the drainage was more than the blocked pipe could handle, which created the puddle.

This is a very good example of how a simple troubleshooting technique solved a problem and made a customer who had doubts about the credibility of the plumbing company happy.

Material Mistakes

Material mistakes account for some strange problems in the DWV systems of buildings. Good plumbers tend to think along the lines of the plumbing code when they are creating a mental picture of a plumbing problem. They basically assume the job was done to code, but this is sometimes far from the truth.

Whether you are troubleshooting a DWV system or any other plumbing problem, don't assume anything. I know it can be hard to think outside of the code, but there are times when you must, as police officers have to think like criminals to catch the bad guys. You have to think like an irresponsible plumber or a rank amateur sometimes to figure out plumbing problems. Let me give you a few examples of what I'm talking about.

The sand trap

This first example could aptly be dubbed the sand trap. The story has to do with a shower drain. A friend of mine who owns his own plumbing business was in my office recently. We were discussing a project that we would be working on together, and he told me this story.

The plumber was called in to work on a shower in the basement bathroom. During the course of his work, he noticed that something was not quite right with the drain in the shower. He removed the strainer plate for a closer look and was amazed at what he saw. Looking through the drain of the shower, he saw sand. His first thought was that the trap had somehow become filled with sand. During his work with the drain, he discovered that it turned freely in the shower base. With a little more investigation, the plumber found that the trap had not been filled with sand. It turns out that the sand was the trap.

Whoever installed the shower never bothered to connect it to a trap or a drain. The fixture simply emptied its waste water into the sand beneath the concrete slab. This is clearly a situation few plumbers would ever imagine.

The building drain that wouldn't hold water

This next case history is about the building drain that wouldn't hold water. The job began when the customer called my office and complained of strong odors in the corner bedroom of the home. It was summer, and I suspected that the customer's problem was an overworked septic field, but I couldn't be sure without an on-site inspection.

I went to the house and met with the homeowner. She started to explain her problem to me, but I could already smell the odor, and I hadn't even gone in the house.

The house was on a pier foundation. I walked around the left side of the home and saw two young children playing in the yard. They wore shorts and no shoes, and were splashing around in a puddle. This seemed odd since it hadn't rained in days.

The closer I got to the back corner of the house, and the children, the worse the smell became. I was starting to get a feeling that made me uncomfortable.

A bathroom had recently been added to the house, near the corner bedroom. Whoever installed the plumbing did so in a way that I'd never seen before. Looking under the house, I could see the problem, but I could hardly believe what I was seeing.

The drainage hanging from the floor joists was piped with schedule 40 PVC. The PVC dropped straight down to the sewer that ran from under the house to the septic tank.

It was the sewer pipe that was creating the problem. You see, someone install slotted drain pipe for the sewer between the house and the septic tank. Much of the sewer had never been buried in the ground. The holes were on top, but the septic tank was full, and raw sewage was seeping out of the slotted sewer pipe. Sewage was puddled under the house, under the bedroom window, and, you guessed it, where the children were playing.

I asked the homeowner who had installed the plumbing, and she claimed that the builder who had built the home some years back had done the work. Can you envision anyone using slotted pipe for a sewer? I couldn't either.

Duct tape

Duct tape is a plumber's friend, and it is capable of many good uses, but it is not meant to be used as a repair clamp on a drainage line.

Many years ago I was called to a house by a homeowner who said he could hear water dripping under his home after flushing the toilet. He thought perhaps the wax seal under his toilet had gone bad. The house was built on a crawlspace foundation. My first step in troubleshooting the situation was to flush the toilet. No water seeped out around the base of the toilet, but I could hear water dripping under the home, just as the homeowner had described.

I took my flashlight and went under the home to inspect the problem. As I worked my way back to the bathroom area, I got a strong whiff of sewage. Once I was close enough to the problem to see the piping and the ground, I noticed a rather large puddle on the ground. Shining my light on the pipe, I could see the remains of duct tape wrapped around the cast-iron drain.

After a closer inspection, I found that the cast-iron pipe had a large hole in the side of it. The pipe appeared to have been hit with a hammer at sometime in the past. Apparently, whoever attempted to repair the hole used duct tape to seal the pipe. Maybe they had done this as a temporary measure until they could get materials to do the job right, I don't know. In any event, the duct tape had long since lost its ability to retain the contents of the pipe, and raw sewage was blowing out the side of the pipe every time the toilet was flushed.

I cut out the damaged pipe and replaced it with a section of ABS and some rubber repair couplings. The repair wasn't a big job, but it was a messy one.

When jobs are not installed with the proper materials, your troubleshooting skills are put to the ultimate test. As plumbers, we are trained to look for normal problems, not problems created by someone's use of illegal materials or installation methods.

If a drain from a lavatory is piped with 1-inch PE pipe, as some I've found have been, it is difficult to understand why the pipe is stopped up or why a snake is so difficult to feed into the drain. Until you see the pipe, it is natural to assume it is of a legal size and material. You cannot, however, assume anything when you're troubleshooting plumbing.

I hope this chapter has opened your eyes to some of the strange calls you may get regarding DWV systems. Furthermore, I hope that by reading this book you have gained new insight to the value of learning effective troubleshooting skills. If you learn and practice the techniques we have talked about, I'm sure your plumbing career will be more successful and satisfying.

This brings us to the end of the book. I wish you good luck in all your endeavors.

Index

Acid neutralizers, 317, 322, 325–326
Acidic water, 87, 320, 322
Adhesive applications (see Caulking)
Aerators, 140, 347
Air blowers, in spas, 101–102, 105
Air chambers, 339–341
Air-volume controls, 284, 295, 296, 297
Anaerobic bacteria, in septic systems,
    311, 312, 313–314
Anti-siphon vacuum breaker, 176
Appliances (see specific appliances)

Bacteria, iron, 319, 322
Ball-type faucets (see Faucets, ball-type)
Ballcocks:
    float-rod, 35–36
    vertical, 36–37
Bar sinks, 59, 143, 163
    faucets, 143, 163
Basin waste pumps, 229, 235
    gravity-type drain, 229
    motors, 229
    running without pumping, 235–236
    uses of, 229
    (See also Sewage ejector pumps)
Basket strainers, in kitchen sinks, 56
    caulking, 56
Bathtubs:
    draining slowly, 77
    escutcheon leaks, 82
    faucets, 143, 156–162, 175, 184–188
    gaskets:
        taper sponge, 81
        tub-shoe, 77
    leaks, 74, 77
    obstructions, in traps and drains, 79–80
    overflow leaks, 81–82

Bathtubs (Cont.):
    putty, 77
    shower heads, 71, 73
    spouts, 71
    traps, 71
    troubleshooting, 71–84
    tub-shoe, 77
    tub spouts, leaks in, 82
    valves, 71, 82–83, 145–148, 156–162,
        175
    vent pipes, blockages in, 81
    waste, 74–75, 77
        lift-and-turn, 75, 78
        mechanical strainer, 75–76, 77,
            78
        rocker type, 76–77, 78
        toe-touch, 75, 78
Bedpan flushers, 262
    diverter flushing fitting, 262
    in water closet, 271
Bibb screws, 136, 138, 139, 149, 150
Bidets, 113–125
    crimped hose, 117
    drain flange, and putty, 120
    faucets, 151–152
        leaks, 123
        options, 124
    gaskets, 121
    hoses, 124
    obstructions, 117
    over-the-rim leaks, 125
    pivot rod, 121
    pop-up plugs, 119, 121, 124–125
    slip-nut, 119
    spray assemblies, 116–119, 123
    spray assembly drain flanges, 121
    supply tubes, 117, 118–119, 122
    threaded tailpiece, 120

ABOUT THE AUTHOR

R. Dodge Woodson has been a licensed master plumber for almost 15 years, with extensive experience in both residential and commercial work. He is the author of the *National Plumbing Codes Handbook, Home Plumbing Illustrated,* and *Plumbing Apprentice Handbook,* all available from McGraw-Hill. Mr. Woodson has also written numerous magazine articles as well as a book on real estate property management.